Selected Papers on Analysis, Probability, and Statistics

American Mathematical Society

TRANSLATIONS

Series 2 • Volume 161

Selected Papers on Analysis, Probability, and Statistics

Katsumi Nomizu
Editor

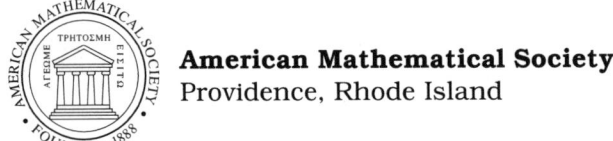

American Mathematical Society
Providence, Rhode Island

1991 *Mathematics Subject Classification.* Primary 30–XX, 31–XX, 35–XX, 42–XX, 47–XX, 60–XX, 62–XX

Library of Congress Cataloging-in-Publication Data
Selected papers on analysis, probability, and statistics/Katsumi Nomizu.
   p. cm. — (American Mathematical Society translations, ISSN 0065-9290; ser. 2, v. 161)
   Includes bibliographical references.
   ISBN 0-8218-7512-4
   1. Mathematical analysis. 2. Probabilities. 3. Mathematical statistics. I. Nomizu, Katsumi, 1924– . II. Series.
QA3.A572   ser. 2 vol. 161
[QA300.5]
515—dc20   94-23002

Copying and reprinting. Individual readers of this publication, and nonprofit libraries acting for them, are permitted to make fair use of the material, such as to copy a chapter for use in teaching or research. Permission is granted to quote brief passages from this publication in reviews, provided the customary acknowledgment of the source is given.

Republication, systematic copying, or multiple reproduction of any material in this publication (including abstracts) is permitted only under license from the American Mathematical Society. Requests for such permission should be addressed to the Manager of Editorial Services, American Mathematical Society, P.O. Box 6248, Providence, Rhode Island 02940-6248. Requests can also be made by e-mail to reprint-permission@math.ams.org.

The owner consents to copying beyond that permitted by Sections 107 or 108 of the U.S. Copyright Law, provided that a fee of $1.00 plus $.25 per page for each copy be paid directly to the Copyright Clearance Center, Inc., 222 Rosewood Drive, Danvers, Massachusetts 01923. When paying this fee please use the code 0065-9290/94 to refer to this publication. This consent does not extend to other kinds of copying, such as copying for general distribution, for advertising or promotional purposes, for creating new collective works, or for resale.

© Copyright 1994 by the American Mathematical Society. All rights reserved.
The American Mathematical Society retains all rights
except those granted to the United States Government.
Printed in the United States of America.

∞ The paper used in this book is acid-free and falls within the guidelines
established to ensure permanence and durability.
♻ Printed on recycled paper.

This volume was typeset using $\mathcal{A}_{\mathcal{M}}\mathcal{S}$-TEX,
the American Mathematical Society's TEX macro system.
10 9 8 7 6 5 4 3 2 1   98 97 96 95 94

Contents

Preface	ix
Nonlinear Partial Differential Equations and Infinite-Dimensional Dynamical Systems HIROSHI MATANO	1
Langevin Equations and Causal Analysis YASUNORI OKABE	19
Analytic Capacity (A Theory of the Szegö Kernel Function) TAKAFUMI MURAI	51
Laplacians on Self-Similar Sets—Analysis on Fractals JUN KIGAMI	75
Statistical Analysis of Mapped Point Patterns—Present Condition of Theory and Application SHIGERU MASE, YOSIHIKO OGATA, and MASAMI TANEMURA	95
Spectra in Random Media SHIN OZAWA	109
The Asymptotic Distributions of Eigenvalues for the Schrödinger Operators with Magnetic Fields HIROYUKI MATSUMOTO	123
Recursive Estimation of Nonparametric Probability Density Functions EIICHI ISOGAI	135

Recent Titles in This Series

161 **Katsumi Nomizu, Editor,** Selected Papers on Analysis, Probability, and Statistics
160 **Katsumi Nomizu, Editor,** Selected Papers on Number Theory, Algebraic Geometry, and Differential Geometry
159 **O. A. Ladyzhenskaya, Editor,** Proceedings of the St. Petersburg Mathematical Society, Volume II
158 **A. K. Kelmans, Editor,** Selected Topics in Discrete Mathematics: Proceedings of the Moscow Discrete Mathematics Seminar, 1972–1990
157 **M. Sh. Birman, Editor,** Wave Propagation. Scattering Theory
156 **V. N. Gerasimov, N. G. Nesterenko, and A. I. Valitskas,** Three Papers on Algebras and Their Representations
155 **O. A. Ladyzhenskaya and A. M. Vershik, Editors,** Proceedings of the St. Petersburg Mathematical Society, Volume I
154 **V. A. Artamonov et al.,** Selected Papers in K-Theory
153 **S. G. Gindikin, Editor,** Singularity Theory and Some Problems of Functional Analysis
152 **H. Draškovičová et al.,** Ordered Sets and Lattices II
151 **I. A. Aleksandrov, L. A. Bokut′, and Yu. G. Reshetnyak, Editors,** Second Siberian Winter School "Algebra and Analysis"
150 **S. G. Gindikin, Editor,** Spectral Theory of Operators
149 **V. S. Afraĭmovich et al.,** Thirteen Papers in Algebra, Functional Analysis, Topology, and Probability, Translated from the Russian
148 **A. D. Aleksandrov, O. V. Belegradek, L. A. Bokut′, and Yu. L. Ershov, Editors,** First Siberian Winter School "Algebra and Analysis"
147 **I. G. Bashmakova et al.,** Nine Papers from the International Congress of Mathematicians, 1986
146 **L. A. Aĭzenberg et al.,** Fifteen Papers in Complex Analysis
145 **S. G. Dalalyan et al.,** Eight Papers Translated from the Russian
144 **S. D. Berman et al.,** Thirteen Papers Translated from the Russian
143 **V. A. Belonogov et al.,** Eight Papers Translated from the Russian
142 **M. B. Abalovich et al.,** Ten Papers Translated from the Russian
141 **H. Draškovičová et al.,** Ordered Sets and Lattices
140 **V. I. Bernik et al.,** Eleven Papers Translated from the Russian
139 **A. Ya. Aĭzenshtat et al.,** Nineteen Papers on Algebraic Semigroups
138 **I. V. Kovalishina and V. P. Potapov,** Seven Papers Translated from the Russian
137 **V. I. Arnol′d et al.,** Fourteen Papers Translated from the Russian
136 **L. A. Aksent′ev et al.,** Fourteen Papers Translated from the Russian
135 **S. N. Artemov et al.,** Six Papers in Logic
134 **A. Ya. Aĭzenshtat et al.,** Fourteen Papers Translated from the Russian
133 **R. R. Suncheleev et al.,** Thirteen Papers in Analysis
132 **I. G. Dmitriev et al.,** Thirteen Papers in Algebra
131 **V. A. Zmorovich et al.,** Ten Papers in Analysis
130 **M. M. Lavrent′ev, K. G. Reznitskaya, and V. G. Yakhno,** One-dimensional Inverse Problems of Mathematical Physics
129 **S. Ya. Khavinson,** Two Papers on Extremal Problems in Complex Analysis
128 **I. K. Zhuk et al.,** Thirteen Papers in Algebra and Number Theory
127 **P. L. Shabalin et al.,** Eleven Papers in Analysis
126 **S. A. Akhmedov et al.,** Eleven Papers on Differential Equations
125 **D. V. Anosov et al.,** Seven Papers in Applied Mathematics
124 **B. P. Allakhverdiev et al.,** Fifteen Papers on Functional Analysis
123 **V. G. Maz′ya et al.,** Elliptic Boundary Value Problems
122 **N. U. Arakelyan et al.,** Ten Papers on Complex Analysis

(*Continued in the back of this publication*)

Preface

This is a collection of several papers that originally appeared in the journal Sugaku in Japanese. These translated articles would normally appear in the AMS journal **Sugaku Expositions**. In order to expedite publication, the AMS has chosen, with the consent of the Mathematical Society of Japan, to publish them as a volume of selected papers in the Society's Translations Series 2.

This volume contains papers in the general area of mathematical analysis as it pertains to probability and statistics, dynamical systems, differential equations, and analytic function theory. The papers by Okabe; Mase, Ogata, and Tanemura; and Isogai are in the areas of probability and statistics. Okabe's paper involves stochastic differential equations. The papers by Ozawa and Matsumoto are concerned with spectra of differential operators. In particular the spectra of the Laplacian and the Schrödinger operator respectively. The paper by Matano studies nonlinear partial differential equations which generate dissipative dynamical systems. Fractal analysis on self-similar sets is the topic of the paper by Kigami while the paper by Murai is concerned with the global structure of analytic functions.

Nonlinear Partial Differential Equations and Infinite-Dimensional Dynamical Systems

Hiroshi Matano

Prologue

The origin of the modern theory of dynamical systems can be traced back to H. Poincaré. The theory has made tremendous progress in this century. It is natural to regard this theory as an organism composed of many disciplines, including topology, ergodic theory, and analysis.

The scope of the theory of dynamical systems has extended from finite-dimensional cases to infinite-dimensional cases. For example, it is known that variational problems, fluid dynamics, soliton theory, and functional differential equations are formulated in the framework of an infinite-dimensional dynamical system. Each field has its own background and methodology; nevertheless, the introduction of the viewpoint of dynamical systems has brought us a deep understanding in a unified way.

Recently, the effectiveness of the theory of dynamical systems in nonlinear diffusion equations has attracted wide attention. Although there was no lack of a "qualitative theory" viewpoint in this field, it previously played only a formal role to provide a general mathematical framework. The reason for this is that a mathematical language had not yet developed enough to describe a variety of complex phenomena such as the creation and annihilation of patterns. Many powerful mathematical tools have since been developed in this field. Methods of bifurcation and singular perturbation are typical examples that helped to explore new aspects of the phenomena. The deeper the understanding of the phenomena has become, the better recognized is the importance of dynamical point of view.

A new trend, related to an application of the theory of systems, appears in the study of elliptic PDE's; namely the behaviors of their solutions are investigated by regarding solutions as orbits of the associated dynamical system. As is well known, the initial value problem for elliptic PDE's is not well-posed, and hence, there are, in general, no solutions. It may sound strange to try to put an ill-posed problem in the framework of dynamical systems; however, for some classes of problems, this treatment becomes very powerful. We present in §3 the author's recent work in this direction.

1991 *Mathematics Subject Classification*. Primary 35B05.
This article originally appeared in Japanese in Sûgaku **42** (4) (1990), 289–303.

One of the important classes of infinite-dimensional dynamical systems is called "dissipative systems". This is not exactly the counterpart of energy dissipative systems in physics; however, qualitatively they are similar to some extent. Most PDE's of diffusion type and the elliptic equations mentioned above belong to this class, along with some classes of equations in fluid dynamics. The characteristic feature of dissipative systems is the existence of an attractor of finite Hausdorff dimension. In this paper we focus mainly on dynamical systems of dissipative type and try to expose the role of the theory of dynamical systems for the analysis of PDE's from the author's point of view.

The fields related to dynamical systems are so broad that we can present here only the tip of an iceberg of results centered around the interaction between nonlinear analysis and dynamical systems. For instance, I will not touch on Hamiltonian systems, which is another important class of dynamical systems, nor on the topics of chaos and turbulence. Of course, this does not mean that they lack importance in the theory of dynamical systems; it simply indicates how the author has committed himself to dynamical system theory so far. The problems of chaos and turbulence are themselves grand themes, and the author would like to tackle them in the future. The progress in the theory of dynamical systems is very rapid and extends its hands to various fields. We anticipate that it will give us a powerful viewpoint for the analysis of nonlinear phenomena. We hope this article provides a a flavor of the vivid theory of dynamical systems.

1. Dynamical system theory and related fields

1.1. Dawn of the qualitative theory. "The qualitative theory of differential equations" was born as antithesis to the methodology of differential equations which had been strongly tied to quadrature. Henri Poincaré (1854–1912) first explored this field through a succession of four papers published from 1881 to 1886 (see [**23, 24**]; both are contained in the first volume of his collected works). He mainly treated two-dimensional ODE systems (three-dimensional in the fourth paper). He discussed and classified the long-term behaviors of solutions by using the techniques which were the precursors of modern topological or ergodic methods. In this manner he pointed out the promising direction of "qualitative theory" by showing various "properties" of solutions without solving the differential equations explicitly. His most striking and decisive paper, in 1889, which accelerated the change of the paradigm, is on the nonintegrability of the three-body problem (the Bruns-Poincaré Theorem).

The three-body problem is a well-known classical problem in celestial mechanics, which describes the motion of three point masses in space governed by Newton's law. The two-body problem had already been solved by Newton himself, and Kepler's law that the orbits of planets are ellipses was proved to be true. However, if the number of point masses is greater than two, then the problem becomes much more difficult and many attempts to solve it had failed. Here "solve" means to find solutions by quadrature, which is equivalent to finding invariants—called "integrals"—equal in number to the number of unknowns. Poincaré proved the nonexistence of independent integrals besides the classical one and gave the last word on the possibility of solving the three-body problem by quadrature. He also posed a question about the validity of the perturbation method by showing that the series, which appears in the study of the stability of the solar system, does not converge uniformly (a half century later, the difficulty he pointed out was overcome, and the perturbation method recovered its validity). These conclusions indeed brought a serious crisis to the basis of analytical mechanics, but at the same time it was an opening of a new era in the qualitative theory of analytical mechanics.

1.2. Definition of dynamical system. The concept of "dynamical system" is basic to the construction of the qualitative theory of differential equations. This concept focuses on the transformation of phase space that is associated with equations instead of treating them directly. A *dynamical system* (or *flow*) is a family of transformations naturally arising from an initial value problem of an autonomous differential equation $du/dt = f(u)$ that is assured of unique, and global existence of solutions for $-\infty < t < \infty$. For a more precise definition see [26]–[28]. However, the above notion of "flow" is too strong for our purpose; namely, the equations describing an irreversible process such as diffusion equations or a viscous fluid model, in general, cannot be solved in the direction of negative time (i.e., nonexistence of solutions), or even if they exist, solutions do not depend continuously on initial data, i.e., not "well-posed" in the negative time direction. This requires an extended notion called "semiflow" to handle the equations that are well-posed only for the positive time direction.

Let X be a topological space, and let $\mathbf{R}^+ = [0, \infty)$. A mapping $\phi \colon \mathbf{R}^+ \times \mathbf{X} \to \mathbf{X}$ is called a *semiflow* or *semidynamical system* with X being a *phase space* when the following three conditions are satisfied:

(a) ϕ is continuous.
(b) $\phi^0 = x$ (for each $x \in X$).
(c) $\phi^t \circ \phi^s = \phi^{t+s}$ (for each $t, s \in \mathbf{R}^+$),

where $\phi^t \colon X \to X$ represents a transformation of X when x is regarded to be an independent variable for a fixed t (i.e., $\phi^t(x) = \phi(t, x)$).

In what follows we focus only on the study of semiflows. A flow is, of course, a special case of a semiflow. The notion of C_0-semiflow in evolutionary theory is very close to that of a semiflow; in fact, sometimes there is no distinction between them. A "local semiflow", which allows solutions to blow up, is an extended notion of semiflow, and the domain of a local semiflow is a subset of $\mathbf{R}^+ \times \mathbf{X}$. Sometimes a local semiflow is simply called a semiflow.

DEFINITION 1.1. Let ϕ be a semiflow on X.

(1) $O^+(x) \stackrel{\text{def}}{=} \{\phi^t(x) \mid t \geq 0\}$ is called *the positive semiorbit* through x.

(2) A curve $\{p(t) \mid -\infty < t < \infty\}$ is called *the total orbit* through x when $p \colon \mathbf{R} \to \mathbf{X}$ satisfies $\phi^t(p(s)) = p(t+s)$ (for each $t \geq 0$ and $s \in \mathbf{R}$) and $p(0) = x$.

(3) A set $S \subset X$ is called *invariant in the positive direction* (or *positively invariant*) when $\phi^t(S) \subset S$ (for each $t \geq 0$) is satisfied.

(4) S is called *invariant* if for an arbitrary point x in S there exists a total orbit contained in S through x (this is equivalent to $\phi^t(S) = S$ (for each $t \geq 0$)).

(5) $\omega(x) \stackrel{\text{def}}{=} \bigcap_{t \geq 0} \text{closure}(O^+(\phi^t(x)))$ is called the *ω-limit set* of x.

Limit sets in the negative time direction are similarly defined, but they are omitted here for notational simplicity; in fact, the orbit in the negative direction may exist but not in a unique way and in general branches off into infinitely many orbits. The total orbit through x, therefore, is not uniquely determined. It is known (see [5, 15]) that uniqueness of the solution in the negative direction (inverse uniqueness) holds for parabolic equations, including diffusion and viscous fluid equations. In this case, there exists at most one orbit passing through a point x. For definitions of terms such as *equilibrium points, closed orbit, stable manifold,* and *unstable manifold* the reader is referred to [6]–[8] and [25]–[28].

1.3. Dimension of attractor.
Motions, such as Brownian motion and the flow of water with infinite degrees of freedom can be characterized as semiflows in infinite-dimensional space. An important class, called dissipative systems, in this category, however, behaves like finite-dimensional systems in the long run. This phenomen was well known empirically among physicists through Fourier analysis, for instance, but mathematically its origin can be traced back to Kolmogorov's paper [11] in 1941. Complete rigorous proofs came in the 1970's: partial results for two-dimensional viscous flows by O. A. Ladyzhenskaya, and later more general results by J. Mallet-Paret (see [12, 13]). In mathematical terminology, the above results can be expressed by saying the Hausdorff dimension of the attractor is finite.

In what follows, a phase space X is always assumed to be a complete metric space.

DEFINITION 1.2. A semiflow ϕ on X is said to be a *dissipative system* (or *point dissipative system*) if the following two conditions are satisfied:

(a) ϕ is a *compact* semiflow; namely, for any $t > 0$ and bounded set $B \subset X$, $\phi^t(B)$ is relatively compact.

(b) There exists a bounded set B_0 such that for any $x \in X$, $\phi^t(x) \in B_0$ is satisfied for all large $t \geq 0$.

The concept of dissipative system defined above is not quite the same as the energy dissipative system in physics; however, if thermodynamically irreversible processes such as viscosity, friction, and diffusion are involved in the systems, most of them become dissipative systems in the above sense. Note that the requirement of compactness in the above condition (a) can be weakened to some extent.

NOTATION 1.3.

$$\mathscr{A} = \{x \in X| \text{ there exists a total orbit passing through } x\}.$$

PROPOSITION 1.4. *Suppose ϕ is a dissipative system. Then*

(i) $\mathscr{A}$ *is a nonempty compact invariant set and is maximal among compact invariant sets.*

(ii) *For an arbitrary bounded set B,*

$$\lim_{t \to \infty} \sup_{x \in B} \text{dist}(\phi^t(x), \mathscr{A}) = 0,$$

where "dist" stands for the distance between two sets.

We call set A satisfying properties (i) and (ii) *the global attractor*. The global attractor contains a limit set of an arbitrary point. Also, it is easily checked that the global attractor contains all equilibrium points, periodic orbits, and their unstable manifolds. It follows from Proposition 1.4 that a dissipative system always has a global attractor. The following result is known as the finite dimension theorem for global attractors.

THEOREM 1.5 (Mallet-Paret). *Let X be a Banach space, and let ϕ be a semiflow on X which is assumed to be dissipative and of C^1-class. Then the Hausdorff dimension of the global attractor $\mathscr{A}$ is finite.*

1.4. Inertial manifold.
The idea of "inertial manifold" came from efforts to find a finite-dimensional system which approximates the dynamics of a dissipative system. The name inertial manifold was given by C. Foias.

A subset M of a phase space X is called *the inertial manifold* of a semiflow ϕ if M satisfies the following conditions:

(M1) M is a finite-dimensional C^1 (or Lipschitz) manifold.

(M2) M is positively invariant; i.e., $\phi^g(M) \subset M$ (for each $t \geq 0$).

(M3) M is stable.

(M4) M has global attractivity; i.e., for arbitrary $x \in X, \lim_{t \to \infty} \text{dist}(\phi^t(x), M) = 0$.

It is sometimes required that M be normally hyperbolic, which means that the attractivity of M is quite strong, at least near M. (M2)–(M4) imply that M contains the global attractor $\mathscr{A}$. From (M1) and (M2) we can see that the restriction of ϕ to M becomes a finite-dimensional semiflow, and hence, it is anticipated that the semiflow can be explicitly represented by finitely many ordinary differential equations. Suppose that this ODE system can be specified. Then important information on the structure of the global attractor $\mathscr{A}$ follows, as well as the properties of the flow near it.

The study of inertial manifolds is being actively pursued mainly in France and in the U.S.A. (see [3, 14, 25]). Since it is still in embryo, most of the works are devoted to finding a nice class of equations where inertial manifolds exist, and we have not yet reached the stage of specifying the ODE systems on them. Inertial manifolds are, in general, obtained indirectly through existence theorems, and hence, it is not an easy task to specify the locations, much less to study the structure of semiflows on them. One sufficient condition for the existence of inertial manifolds is given by the spectrum of the differential operator contained in the model. A rigorous treatment of inertial manifolds has just started recently; however, the original idea can be traced to Kolmogorov's work [11].

Independently of the above stream of work in the U.S.A. and in France, which focused on the general framework, more concrete reaction diffusion systems are being studied in Japan in order to learn the detailed properties of inertial manifolds. Concrete forms of differential equations on inertial manifolds have been obtained by S.-I. Ei, M. Mimura, and Y. Morita (see [4, 20]), although they treat special equations.

1.5. Structural stability. A. A. Andronov and L. S. Pontrjagin introduced the notion of structural stability in 1937. A system is said to be *structurally stable* if the structure as a dynamical system remains the same when a small perturbation is added to it (for instance by changing the coefficients of the equations). Most of the physical phenomena observed in nature are regarded as structurally stable.

Morse-Smale systems form a typical class of structurally stable systems.

DEFINITION 1.6. A differentiable semiflow ϕ is called a *Morse-Smale system* (M-S system) if it is a dissipative system and satisfies the following properties:

(a) The nonwandering set consists of finitely many equilibrium points and periodic orbits, and they are all hyperbolic.

(b) Stable and unstable manifolds of each equilibrium point or periodic point intersect transversally.

(c) Each unstable manifold is of finite dimension.

(d) Both ϕ^t and $D_x\phi^t$ are injective on the global attractor $\mathscr{A}$ (for each fixed t).

The nonwandering set is the set of nonwandering points, and a point $x \in X$ is said to be a nonwandering point if $U \cap O^+(\phi^t(U)) \neq \varnothing$ for any neighbourhood U of x and any $t > 0$. For finite-dimensional flows, conditions (c) and (d) are obviously satisfied; nor is it essential to assume that they are dissipative. It can be proved that M-S systems form a dense open set in two-dimensional dynamical systems; however, it does not hold for higher dimensions. Nevertheless, there are several important classes of M-S type.

Finite-dimensional M-S systems are structurally stable, while infinite-dimensional M-S systems are not, in general; however, their attractors become structurally stable. Namely, we have

THEOREM 1.7 (Oliva [7]). *An infinite-dimensional M-S system is $\mathscr{A}$-stable, that is, the structure of its global attractor $\mathscr{A}$ does not change when the system is slightly perturbed.*

Even if a system is not of M-S type, its attractor has some sort of continuity or semicontinuity occasionally. Making use of general results coming from the above discussions and similar techniques as in the theory of inertial manifolds, many singular perturbation problems have been considered. For instance, the relation between the structures of attractors of the semilinear wave equation with damping (for small ε),

$$(1) \qquad \varepsilon^2 u_{tt} + u_t - \Delta u - f(u) = 0,$$

and its limiting ($\varepsilon = 0$) parabolic equation,

$$(2) \qquad u_t - \Delta u - f(u) = 0,$$

can be handled, and the behaviors of viscous fluid in a very thin three-dimensional domain and the associated two-dimensional flow can be compared.

2. Nonlinear diffusion equations

2.1. Diffusion phenomena and pattern formulation.

While visiting the University of Arizona for a couple of months in 1981, the author attended a lecture by the famous mathematical biologist A. T. Winfree. He distributed to the audience vinyl sheets a couple of inches square. There were strange patterns of stripes when one looked through the sheet to the light. The shape of the patterns was changed by pressing the sheet with a finger. It was soon noticed that, without pressing the sheet, the many fine patterns of concentric or spiral shapes were moving around like living creatures and changing their forms. We could not stop watching that mysterious phenomenon. See Figure 1.

The author learned later that the sheet was made of two vinyl sheets with a special chemical species in between and that the chemical reaction called the Belousov-Zhabotinsky (B-Z) reaction was being demonstrated. All the author knew about chemistry then was a formula $A + B \rightleftharpoons C$ from a high-school textbook. So vivid and mysterious a chemical reaction was really striking.

Most of the phenomena, including the diffusion of particles such as in the chemical reaction described above, are known to be modelled by nonlinear diffusion equations; in particular, those which model chemical reactions are called "reaction diffusion equations". For the scalar case, the model takes the form

$$(3) \qquad u_t = \Delta u + f(u) \qquad \left(\Delta = \frac{\partial^2}{\partial x_1^2} + \cdots + \frac{\partial^2}{\partial x_n^2}\right),$$

where Δu represents the diffusive effect and f is a nonlinear term corresponding to the reaction. For the general case, the model becomes a system of equations of this type.

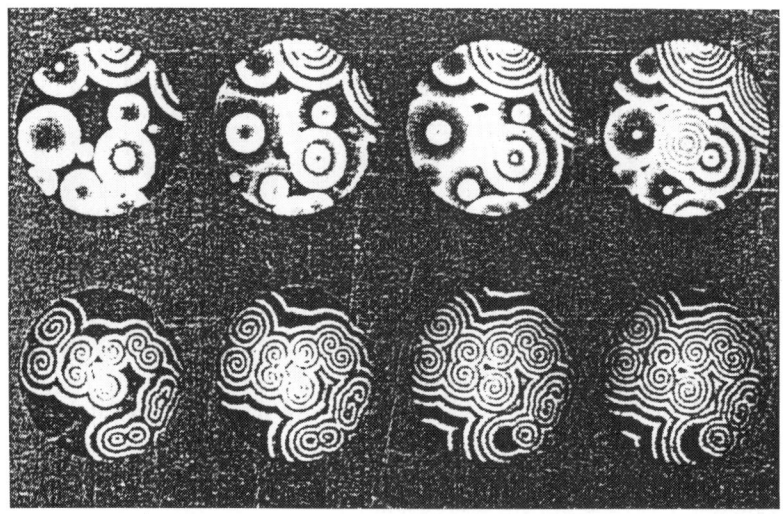

FIGURE 1. Pattern formation in a chemical reaction.

Nonlinear diffusion phenomena are studied in many other fields, such as physics, chemistry, and biology. Even when restricted to the mathematical aspects, there are considerably many researchers working in this field; however, it is rather recently that people have started to do research on dynamic pattern formation problems such as the B-Z reaction.

Since space is limited, we consider here only several topics related to dynamical systems from the broad theme of nonlinear diffusion phenomena, and we give the reader a bird's-eye view.

2.2. Dynamical systems determined by nonlinear diffusion systems. We can generalize equation (3) to an abstract parabolic equation in an appropriate function space X as

$$\frac{du}{dt} = Au + f(u), \tag{4}$$

where A is a generator of an analytic semigroup and f is a smooth mapping from X (or a dense subspace of X) into X. It is well known under these conditions that (4) generates a semiflow on X (or a dense subspace) (see Henry [8]). The most typical case is that $A = \Delta$ (Laplace operator). In such a case it is well known that A generates the analytic semigroup when X is taken to be L^p—or the usual Sobolev space; moreover, Kyuya Masuda proved in 1970 that this is also true even when X is a space of continuous functions. Which function space is best as a phase space depends on the character of $f(u)$ and the purpose of the analysis.

If A has a resolvent which is compact and f satisfies appropriate conditions, then the semiflow generated by (4) becomes a dissipative system. Most of the nonlinear diffusion equations defined on a bounded domain belong to this category.

2.3. Properties of semiflow. There are special semiflows determined by nonlinear diffusion equations that satisfy the following properties:

(P1) ϕ is a strongly order-preserving system;
(P2) ϕ is a Morse-Smale system.

(P1) holds for most scalar equations and for several classes of systems, such as cooperative ones and competitive systems of two species. "Strongly order-preserving system" means a system such that the associated semiflow defined on a metric space with order structure satisfies the usual order-preserving property

$$x \geq y \Rightarrow \phi^t(x) \geq \phi^t(y) \quad (\text{for each } \geq 0)$$

in a slightly strong sense. The general framework for this was established by M. W. Hirsch and the author independently in the 1980's (see [**10, 18**]). The system has several remarkable properties, which are not shared with other dynamical systems; for example, closed orbits are necessarily unstable and there always exists a stable equilibrium point in a relatively compact stable positively invariant set. The theory is applied to various equations, besides nonlinear diffusion equations, to analyse the structure of solutions. (One can find applications of the theory of similar dynamical systems to nonlinear diffusion equations in [**2**] and [**17**].)

The Morse-Smale property (P2) holds, for instance, for a single nonlinear diffusion equation in a one-dimensional space

$$u_t = u_{xx} + f(u),$$

where we assume that each equilibrium point is of hyperbolic type. (It always holds, without this assumption, that stable and unstable manifolds intersect each other transversally!) This was proved in the mid-1980's by D. Henry and S. Angenent independently [**1, 9**]. The main idea of the proof is that the number of zero points of a linear parabolic equation of second order does not increase in time, which is inherent in the one-dimensionality of space. Incidentally, this "nonincreasing property of zero points" was known a rather long time ago and was "rediscovered" several times in the past; however, the author in 1978 (see [**16**]) was the first to make it clear that it is quite useful for the analysis of the qualitative properties of semiflows.

2.4. Minimal surface and pattern formation. The fact that some parameter contained in a system is quite small plays an important role for the study of the pattern formation problem. For simplicity we first focus on the single equation in a domain Ω in $\mathbf{R}^n$ with a small parameter ε added to (3):

$$(5) \qquad \varepsilon u_t = \varepsilon^2 \Delta u + f(u).$$

We assume that $f(u) = u(1 - u^2)$, which appears quite often in applications. When the parameter $\varepsilon > 0$ is sufficiently small, (5) can be formally approximated by the equation $f(u) = 0$, which contains no derivatives of u. A problem such as (5) where a small parameter appears in the coefficient of the highest-order derivatives of the differential equations is called *a singular perturbation problem*. It is not difficult to imagine that a solution $u(x,t)$ of (5) rapidly approaches a step function with values either $+1$ or -1 in a short time unless it starts from exceptional initial data. Of course, this is an approximation since $u(x,t)$ must be smooth for all time; nevertheless it is quite sharp and useful.

At each time, u changes rapidly from -1 to $+1$ at the boundary surface (called the "interface") between the regions $u > 0$ and $u < 0$. This region is called the "internal transition layer", the width of which is proportional to ε and encloses the interface. It is clear that if we know how the interface moves, then we can grasp the motion of the solution u, since u is approximately equal to the constant (± 1) away from the interface. Roughly speaking, the motion of the interface tends to the direction which makes the area of it as small as possible; namely, the normal velocity at each point of the interface is proportional to the mean curvature there. It is not difficult to derive this formally if we consider equation (5) in the framework of a variational formula; however, a rigorous treatment is quite difficult, and recent developments of the singular perturbation method have made it possible.

From the viewpoint of the theory of dynamical systems, the situation described in the above may be recast as follows. We let ϕ_ε be the semiflow determined by (5). The phase space of this semiflow is a usual function space and should be considered for the moment separately from the interfacial dynamics in the above. As the parameter gets smaller, there appears an invariant manifold M in the phase space X of ϕ_ε which strongly attracts all the points nearby. Once an orbit comes close enough to M, it evolves along M according to the dynamics imposed on M by the semiflow. Therefore, only the dynamics on the invariant manifold is visible to an observer. Moreover, each point of M is very close to a step-like function (with sharp interfaces), and thus, the dynamism of interfacial curvature-motions is concentrated on the invariant manifold M.

There are, of course, several problematical points in the above scenario. The actual motion of interfaces (namely the behavior of $\phi_\varepsilon | M$) not only evolves according to mean curvature motion but is also influenced by an extremely weak interaction between interfaces. Due to various reasons, such as the one just described, the scenario in the last paragraph may not be valid as it stands. We believe, however, that an appropriately modified version of it is correct. This scenario, if correct, predicts that interfaces either disappear or approach approximate minimal surfaces as the time t goes to infinity. This is in good agreement with the results observed in numerical simulations.

In more general systems of reaction-diffusion equations, although the situation becomes more involved because the number of unknowns increasis and the reaction term is replaced by general ones, we still believe that the scenario described in the above is more or less valid. There are many interesting questions such as: What types of interfacial configurations are stable? In what situation does a Hopf bifurcation give rise to an oscillation of the interface? Naturally, these questions ought to be considered as stability analysis and bifurcation problems in singular perturbation limits. Analyses of such problems are very difficult. The SLEP (Singular Limit Eigenvalue Problem method) recently developed by Y. Nishiura, M. Mimura, and H. Fujii seems to be promising as a powerful tool to attack the problems in singular limits, although only one space dimension problems can be handled at present. See [21, 22], and [29]. (Reference [29] was added by the translator.)

3. Elliptic partial differential equations and dynamical systems

3.1. Background of the problem. As is well known, initial value problems for elliptic equations are ill posed. That is, either there exists no solution for a given initial condition or, even if a solution exists, the solution does not depend continuously on the initial value. Even though this is the case in general, there are situations in which initial value problems for elliptic equations can be considered, with some tricks, as well posed. Under such circumstances, the theory of dynamical systems provides us with a powerful method to analyse the behavior of solutions and the structure of the solution sets.

Let us now fix a constant $q > 1$ and consider the following semilinear elliptic equation:

$$\text{(6)} \qquad \Delta u - |u|^{q-1}u = 0.$$

Let $a_1, \ldots, a_m$ be a set of arbitrary points of a domain Ω in $\mathbf{R}^n$. If a function $u(x)$ is defined in $\Omega\setminus\{a_1, \ldots, a_m\}$ and satisfies equation (6) there, then one can consider u as a solution of (6) with isolated singularities at $a_1, \ldots, a_m$.

We then try to classify the types of singularities by examining the behavior of the solution u in a neighborhood of the isolated singularities. This is called the "local theory" in the sequel. Equations similar to (6) appear in the Thomas-Fermi theory in quantum mechanics, where the points $a_1, \ldots, a_m$ correspond to the positions of the electrons. In the Thomas-Fermi theory, the function $u(x)$ is restricted to be positive, and there are few possible types of singularities. The objective of "local theory" is to classify all the possible types of singularities that appear in (6) without any restrictions.

Since the latter half of the 1970s, the "local theory" has been developed by Brezis, Lieb, and Véron, and the problem has been settled for large q. On the other hand, as q gets smaller, the number of singularity types increases rapidly, and anisotropic strong singularities (which are very difficult to analyse) appear. Due partly to this complication, there have been very few studies of the latter case. Recently, the author has succeeded in classifying all the types of isolated singularities of (6) completely for two-dimensional space case through joint research with X.-Y. Chen and L. Véron. For details, consult [2]. By the way, we note that the classification in [2] is done based upon the principal part of the expansion of the solutions around a singularity. A detailed study of the higher-order terms in the expansion is being done by T. Ishii.

Let us now consider the special case in which there is only one singularity in the plane:

$$\text{(7)} \qquad \Delta u - |u|^{q-1}u = 0 \quad \text{in } R^2\setminus\{0\}.$$

We call such a solution a "global singular solution", for which the following is applicable.

THEOREM 3.1 (Chen-Matano-Véron). *Let* $u \in C^2(R^2\setminus\{0\})$ *satisfy* (7). *Then the following limits exist*:

$$\text{(8a)} \qquad \omega_0(\theta) = \lim_{r \to 0} r^{2/(q-1)} u(r, \theta),$$

$$\text{(8b)} \qquad \omega_\infty(\theta) = \lim_{r \to \infty} r^{2/(q-1)} u(r, \theta).$$

where (r, θ) *stands for a polar coordinate system. The convergence in* (8) *is with respect to the* $C^2(S^1)$*-topology* $(S = \mathbf{R}/2\pi\mathbf{Z})$. *Moreover, the limit functions* ω_0 *and* ω_∞ *satisfy the ordinary differential equation*

$$\text{(9)} \qquad \frac{d^2\omega}{d\theta^2} + g(\omega) = 0 \quad \text{in } S^1$$

in which $g(s) = (2/(q-1))^2 s - |s|^{q-1}s$.

From now on, we call ω_0 and ω_∞, respectively, "the profile at the singularity" and "the profile at infinity" of the solution $u(x)$. The behavior of a solution is to be classified in terms of these profiles. Now what is the relationship between the two profiles ω_0 and ω_∞? It turns out that they cannot be specified arbitrarily. In order to examine the relationship between the two profiles, let us introduce the energy functional

$$J(\psi) = \int_{S^1} \left\{ \frac{1}{2} \left(\frac{d\psi}{d\theta}\right)^2 - G(\psi) \right\} d\theta$$

(where G is an antiderivative of g). Then an easy computation gives the following proposition.

PROPOSITION 3.2. *Let ω_0 and ω_∞ be determined by* (8). *Then one of the following two alternatives is true.*
(a) $J(\omega_0) > J(\omega_\infty)$;
(b) $\omega_0 = \omega_\infty$ *and* $u(r,\theta) = r^{-2/(q-1)}\omega_0(\theta)$.

The following question naturally comes to our mind. This is a kind of converse statement of Proposition 3.2.

(Q) $\begin{cases} \text{For any pair of solutions } \omega_0 \text{ and } \omega_\infty \text{ of (9) with } J(\omega_0) > J(\omega_\infty), \\ \text{does there exist a solution } u \text{ of (7) satisfying (8)?} \end{cases}$

It is the purpose of this section to answer the question (Q). The answer to this question is "yes". In order to come to this conclusion, it is necessary to study the structure of the solution set of (7). One may call this type of study a "global theory" in contrast to the "local theory" in the above. The result we are going to explain is reported in [19].

3.2. Ellipticity and ill-posedness. As we said before, initial value problems for partial differential equations of elliptic type are ill-posed. We now take a look at this fact from the dynamical system point of view so that the subsequent line of arguments becomes clear.

In order to make the story transparent, let us consider the linearized version of (7):

(10) $$\Delta u = 0 \quad \text{in } \mathbf{R}^2 \setminus \{0\}$$

By changing variables via $t = \log r$ and setting $v = u_t$, one obtains the following equation for $u(t,\theta)$ and $v(t,\theta)$:

$$\frac{d}{dt}\begin{pmatrix} u \\ v \end{pmatrix} = \begin{pmatrix} 0 & 1 \\ -\Delta_s & 0 \end{pmatrix}\begin{pmatrix} u \\ v \end{pmatrix},$$

where $-\Delta_s = -d^2/d\theta^2$ is a differential operator on S^1. This operator is selfadjoint on $L^2(S^1)$ and is positive definite when restricted to $Y = \{u \in L^2(S^1) | \int_{S^1} u \, d\theta = 0\}$. By a suitable transformation on u and v, the last equation can be diagonalized as follows:

(11) $$\frac{d}{dt}\begin{pmatrix} p \\ q \end{pmatrix} = \begin{pmatrix} -\sqrt{-\Delta_s} & 0 \\ 0 & \sqrt{-\Delta_s} \end{pmatrix}\begin{pmatrix} p \\ q \end{pmatrix}$$

This equation is parabolic for p and backward parabolic for q. The former is well-posed in the positive time direction, but is not well-posed in the negative time direction. In fact, $\sqrt{-\Delta_s}$ is an unbounded positive definite operator, and hence $e^{-t\sqrt{-\Delta_s}}$ is not defined for $t < 0$. On the other hand, the latter equation is well-posed in the negative time direction and ill-posed in the positive time direction. Therefore, when one tries to solve the two equations simultaneously, the problem (11) is not well-posed in either direction. In this way, one can understand the reason why an initial value problem for an elliptic equation is ill-posed.

We should, however, notice that equation (11) is well-posed either in the positive or the negative time direction when it is restricted to either a p-subspace or a q-subspace, both of which are infinite dimensional. This fact, although trivial, suggests to us that even an ill-posed evolution equation in some cases can be made well-posed by restricting it to an appropriate invariant submanifold. In fact, the p-subspace and the q-subspace are, respectively, the stable manifold and the unstable manifold of the origin $(p, q) = (0, 0)$ for equation (11).

For the linear equation, the decomposition into p-subspace and q-subspace was a trivial matter. Such a decomposition is not so easy for nonlinear equations. If one can apply to a nonlinear elliptic equation an argument similar to the one in the above, namely, if one can find an appropriate invariant submanifold on which the equation is well-posed, then one has the possibility of defining a semiflow at least on such a submanifold and utilising the theory of dynamical systems to study the solutions of the nonlinear elliptic equation.

Fortunately, for equation (7) one can show that the stable manifold of the set of equilibrium points has fairly nice properties. To explain this, let us rewrite (7) by using $t = \log r$, $v(t, \theta) = r^{2/(q-1)}u(r, \theta)$:

$$(12) \qquad v_{tt} - 2\alpha v_t + v_{\theta\theta} + g(v) = 0 \qquad (t \in \mathbf{R},\ \theta \in S^1)$$

where $\alpha = 2/(q-1)$. This equation defines an "ill-posed flow" in the phase space $(v, v_t) \in H^1(S^1) \times L^2(S^1)$. If we denote by $\mathscr{E}^*$ the totality of the equilibrium points of (12), denote by $W^s(\mathscr{E}^*)$ its stable manifold, set $V = H^1(S^1) \times \{0\}$, and denote by P the natural projection from the whole space onto V, then we can show the following.

PROPOSITION 3.3. *P maps $W^s(\mathscr{E}^*)$ onto V homeomorphically. Therefore, the semiflow defined on $W^s(\mathscr{E}^*)$ by (12) can be identified with the semiflow on $V \simeq H^1(S^1)$. By letting $\mathscr{E}$ denote the set of all solutions of (8), namely,*

$$\mathscr{E} = \left\{ \omega \in C^2(S^1) \,\bigg|\, \frac{d^2\omega}{d\theta^2} + g(\omega) = 0 \text{ in } S^1 \right\}$$

then one easily finds that $\mathscr{E}^* = \{(\omega, 0) | \omega \in \mathscr{E}\}$, and $P\mathscr{E}^* = \mathscr{E}$. This shows that the set of equilibrium points of the semiflow (which we denote by ϕ) on V, obtained in the last proposition, is equal to $\mathscr{E}$.

Theorem 3.1 implies that global solutions of (7) converge to a point of $\mathscr{E}$ as $t \to \infty$. This means that (v, v_t) is on $W^s(\mathscr{E}^*)$. Therefore, the behavior of the solution $v(t)$ can be described completely by the semiflow ϕ. Thus, *the problem of finding the global solutions of (7) is reduced to determining all the orbits of the semiflow ϕ*. See Figure 2.

Proposition 3.3 can be proven by using the fact that the exterior problem

$$(13) \qquad \begin{aligned} \Delta u - |u|^{q-1} &= 0 & (|x| > 1), \\ u &= \psi & (|x| = 1), \end{aligned}$$

has a unique solution for an arbitrarily given boundary value $\psi \in C(S^1)$ and that the solution depends on the boundary value ψ continuously. The maximum principle is used to prove this fact. In dealing with more general equations, one cannot expect

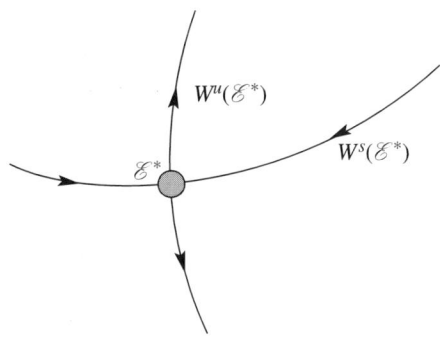

FIGURE 2. A schematic picture of the flow of the equation of elliptic type. Both W^u and W^s are of infinite dimension, although W^u is finite-dimensional for parabolic equations. Orbits cannot be defined at almost all points outside the two invariant manifolds.

such a lucky situation as this, and there is no guarantee that the stable manifold $W^s(\mathscr{E}^*)$ can be expressed as a graph over V. However, in many cases it is possible to construct $W^s(\mathscr{E}^*)$ locally near $\mathscr{E}^*$. Even this alone provides us with substantial information on the structure of the solutions to the original elliptic equation. We will touch upon this in §3.4.

3.3. Structure of the semiflow. The semiflow ϕ given in Proposition 3.3 can be considered as being generated by the initial value problem

$$
\begin{aligned}
&v_{tt} - 2\alpha v_t + v_{\theta\theta} + g(v) = 0 \quad (t > 0,\ \theta \in S^1), \\
&v(0, \cdot) = \psi,
\end{aligned}
\tag{14}
$$

which can be obtained from (13) via a change of variables. Although this problem is a boundary value problem, we, on purpose, regard this as an initial value problem since it has a unique solution for a given ψ. One can in fact rewrite equation (14) as a genuine parabolic equation. If we set

$$A_1 = \alpha - \sqrt{\alpha^2 - \Delta_s}, \qquad A_2 = \alpha + \sqrt{\alpha^2 - \Delta_s}$$

(where $\Delta_s = d^2/d\theta^2$, $\alpha = 2/(q-1)$), then (14) is equivalent to

$$
\begin{aligned}
&\frac{dv}{dt} = A_1 v + f(v) \quad (t > 0), \\
&v(0) = \psi
\end{aligned}
\tag{15}
$$

in which $f(w)$ is given by

$$f(w) = \int_0^\infty e^{-\tau A_2} g(\phi^\tau(w))\, d\tau.$$

In this way, one can rewrite an elliptic partial differential equation as an abstract parabolic equation only because the solution (v, v_t) is known to be on the stable manifold $W^s(\mathscr{E}^*)$. One can read off from the expression (15) that ϕ is a $C^{1+\gamma}$ class semiflow, where $0 < \gamma < 1$ is a constant dependent on q.

In order to investigate the semiflow ϕ, one needs first of all to know the structure of the set $\mathscr{E}$ of equilibrium points. Simple reasoning shows that $\mathscr{E}$ has $k_0 + 3$ connected components (where k_0 is the largest integer less than $2/(q-1)$).

$$\mathscr{E} = \mathscr{E}^0 \cup \mathscr{E}^+ \cup \mathscr{E}^- \cup \mathscr{E}_1 \cup \mathscr{E}_2 \cup \cdots \cup \mathscr{E}_{k_0}$$

Here $\mathscr{E}^0 = \{0\}$ and $\mathscr{E}^\pm\{\pm\alpha^\alpha\}$ correspond to constant solutions of (14), and $\mathscr{E}_k$ is the set of nonconstant $2\pi/k$ periodic solutions on S^1. Since equation (9) is equivariant with respect to the rotations on S^1, each $\mathscr{E}_k$ is homeomorphic to S^1. In this way, $\mathscr{E}$ turns out to be a set consisting of three isolated points and k_0 closed curves. See Figure 3.

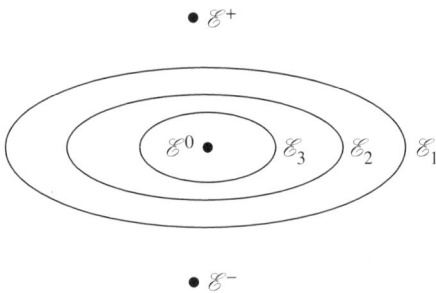

FIGURE 3. A schematic of the set of equilibrium points.

Now we can summarize the relation between the global solutions of (12) and the orbits of the semiflow ϕ as follows:

(i) The curve followed by a global solution of (12) corresponds to a full orbit of the semiflow ϕ. The converse of this is also valid.

(ii) Therefore, if global solutions of (12) are identified with their values at $t = 0$, then there is a one-to-one correspondence between the global solutions and the full orbits.

(iii) Theorem 3.1 says that each full orbit is bounded, and hence, the union of all the full orbits agrees with the global attractor $\mathscr{A}$.

From (ii) and (iii) in the above, one can conclude that *the set of all global singular solutions of* (7) *can be identified with the global attractor* $\mathscr{A}$.

Theorem 3.1 together with (i) in the above shows that each full orbit connects two equilibrium points ω_0, ω_∞. Therefore, the question (Q) is now equivalent to the following.

(Q) $\begin{cases} \text{For a given pair of equilibrium points } \omega_0, \omega_\infty, \text{ are there} \\ \text{full orbits connecting these two (from } \omega_0 \text{ to } \omega_\infty)? \\ \text{In other words, is } W^u(\omega_0) \cap W^s(\omega_\infty) \text{ empty or not?} \end{cases}$

Although the details will not be given here, the author has succeeded in giving a complete answer to this question. For example, if k, k' is a pair of integers satisfying $1 \leq k < k' \leq k_0$, then there is no orbit from $\mathscr{E}_k$ to $\mathscr{E}_{k'}$, while there is an orbit connecting an arbitrary point of $\mathscr{E}_{k'}$ to an arbitrary point of $\mathscr{E}_k$. A similar result was known in [2] for the case when k' is an integer multiple of k, but the answer was not known for the general case. If, for example, one applies this general result to the case $k = 2$, $k' = 3$, then one finds that the original problem (7) has a global singular solution which substantially lacks symmetry. In fact, this singular solution is such that it oscillates three times on a small circle centered at the origin of $\mathbf{R}^2$ and oscillates two times on a large circle. Unlike the situation for the usual boundary value problems, here one cannot specify arbitrarily the two profiles at the singularity and at infinity, but rather they are determined naturally from an internal mechanism hidden in the equation. Given this, it seems to be extremely difficult to construct, by the usual analytical methods alone, such a solution which has very few symmetry properties.

The proof of the result in the above is based on the fact that the stable manifold and the unstable manifold of equilibrium points intersect transversally. The latter fact, on the other hand, can be proven by using more or less the same method as that applied to one-dimensional diffusion equations, where we take advantage of the fact that S^1 is one-dimensional.

3.4. Future perspectives

When problem (7) is considered on three-dimensional space ($n = 3$), the corresponding semiflow acts on the phase space which is a function space on S^2. There is no problem in the basic construction of the semiflow which is parallel to the two-dimensional case, but we do not know much about the properties of the semiflow. First of all, the structure of the solution set of the S^2 version of problem (9) still remains obscure. Although it is possible to construct several solutions with various symmetries by using bifurcation theory, one cannot understand the overall structure of the solution set. Therefore, we do not know the structure of the solution set. Therefore, we do not know the structure of the set $\mathscr{E}$ of equilibrium points.

Another difficulty is that we cannot prove the transversality between the stable and the unstable manifolds. The proof in [**19**] essentially uses the one dimensionality of S^1, and it does not generalise to the S^2 case. The situation is the same for the diffusion equation described in §3.2. It is a future task to determine whether or not transversality is valid for equations in higher dimensions. By the way, K. Mischaikov has recently succeeded in gaining detailed information on connecting orbits without using transversality. His result is expected to be applicable to the problem above as well as to pattern formation problems described by fourth-order parabolic equations for which the validity of transversality is not known.

We finally point out that the method of this section can be used to obtain strong results for elliptic equations other than (6). Let us consider the following problem.

$$
(16) \quad \begin{aligned} \Delta u + \lambda e^u &= 0 \quad (|x| < 1), \\ u &= 0 \quad (|x| = 1). \end{aligned}
$$

Here we assume that the space dimension is 3. By a solution of (16) we mean a pair consisting of $u(x)$ and the parameter $\lambda \geq 0$. A theorem by Gidas-Ni-Nirenberg immediately shows that all the classical solutions of (16) are radial; namely, u depends only on $|x|$. Although this is the case, it is unknown whether weak solutions (H^1 solutions) are necessarily radial. At present, $u(x) = 2\log(1/|x|)$, $\lambda = 2$, is the only weak solution known. It is known that the totality of $\mathscr{E}^*$ of the equilibrium points of the ill-posed semiflow determined by the elliptic equation in the above is a three-dimensional manifold homeomorphic to the set of conformal transformations on S^2.

If one can show that the unstable manifold of $\mathscr{E}^*$ is C^1 in a neighborhood of $\mathscr{E}^*$, then one can also prove that there is a nonradial weak solution of (16).

References

1. S. B. Angenent, *The Morse-Smale property for a semilinear parabolic equation*, J. Differential Equations **67** (1986), 212–242.
2. X.-Y. Chen, H. Matrano, and L. Véron, *Anisotropic singularities of solutions of nonlinear elliptic equations in* $\mathbf{R}^2$, J. Funct. Anal. **83** (1989), 50–97.
3. P. Constantin, C. Foias, B. Nicolaenko, and R. Temam, *Spectral barriers and inertial manifolds for dissipative partial differential equations*, J. Dynamics Differential Equations **1** (1989), 45–73.
4. S.-I. Ei and M. Mimura, *Pattern formation in heterogeneous reaction-diffusion-advection systems with an application to population dynamics*, SIAM J. Math. Anal. **21** (1990), 346–361.
5. A. Friedman, *Partial differential equations of parabolic type*, Prentice-Hall, Englewood Cliffs, N.J., 1964.
6. J. K. Hale, *Asymptotic behavior of dissipative systems*, Math. Surveys and Monographs, vol. 25, Amer. Math. Soc., Providence, R.I., 1988.
7. J. K. Hale, L. T. Magalhães, and W. M. Oliva, *An introduction to infinite dimensional dynamical systems—Geometric theory*, Appl. Math. Sci., vol. 47, Springer-Verlag, New York, 1984.
8. D. Henry, *Geometric theory of semilinear parabolic equations*, Lecture Notes in Math., vol. 840, Springer-Verlag, Berlin and New York, 1981.
9. _____, *Some infinite dimensional Morse-Smale systems defined by parabolic differential equations*, J. Differential Equations **58** (1985), 165–205.
10. M. W. Hirsch, *Systems of differential equations that are competitive or cooperative. II: Convergence almost everywhere*, SIAM J. Math. Anal. **16** (1985), 423–439.
11. A. N. Kolmogorov, *The local structure of turbulence in incompressible viscous fluid for very large Reynolds number*, Dokl. Akad. Nauk SSSR **30** (1941), 301–305.
12. O. A. Ladyzhenskaya, *On the dynamical system generated by the Navier-Stokes equations*, J. Soviet Math. **3** (1975), 458-479.
13. J. Mallet-Paret, *Negatively invariant sets of compact maps and an extension of a theorem of Cartwright*, J. Differential Equations **22** (1976), 331–348.
14. J. Mallet-Paret and G. Sell, *Inertial manifolds for reaction-diffusion equations for higher dimensions*, J. Amer. Math. Soc. **1** (1988), 805–866.
15. K. Masuda, *On the analyticity and the unique continuation theorem for the Navier-Stokes equations*, Proc. Japan Acad. **43** (1967), 827–832.
16. H. Matano, *Convergence of solutions of one-dimensional semilinear parabolic equations*, J. Math. Kyoto Univ. **18** (1987), 224–243.
17. _____, *Asymptotic behavior and stability of solutions of semilinear diffusion equations*, Publ. Res. Inst. Math. Sci. **15** (1979), 401–458.
18. _____, *Existence of nontrivial unstable sets for equilibriums of strongly order-preserving systems*, J. Fac. Sci. Univ. Tokyo **30** (1983), 645–673.
19. _____, *Singular solutions of a nonlinear elliptic equation and an infinite dimensional system*, preprint.
20. Y. Morita, *Reaction-diffusion systems in nonconvex domains: Invariant manifolds and reduced form*, J. Dynamics Differential Equations **2** (1990), 69–115.
21. Y. Nishiura and F. Fujii, *Stability of singularly perturbed solutions to systems of reaction-diffusion equations*, SIAM J. Math. Anal. **18** (1987), 1726–1770.
22. Y. Nishiura and M. Mimura, *Layer oscillations in reaction-diffusion systems*, SIAM J. Appl. Math. **49** (1989), 481–514.
23. H. Poincaré, *Mémoire sur les courbes définies par une équation différentielle. I, II*, J. Math. Pures Appl. **3** (1881), 375–422; (1882), 251–296.
24. _____, *Sur les courbes définies par les équations différentielles. I, II*, J. Math. Pures Appl. **4** (1885), 167–244; (1886), 151–217.
25. R. Temam, *Infinite-dimensional dynamical systems in mechanics and physics*, Springer-Verlag, New York, 1988.
26. T. Niwa, *Dynamical systems*, Kinokuniya Series of Mathematical Monographs, vol. 21, Kinokuniya, Tokyo, 1981. (Japanese)

27. T. Saito, *Introduction to analytical mechanics*, Shibundo, 1964. (Japanese)
28. K. Shiraiwa, *Dynamical system theory*, Iwanami Syoten, Tokyo, 1974. (Japanese)
29. Y. Nishiura, *Coexistence of infinitely many stable solutions to reaction diffusion systems in the singular limit*, Dynamics Reported (L. Kirchgraber and H. O. Walther, eds.), Springer-Verlag, in press.

Translated by YASUMASA NISHIURA

Langevin Equations and Causal Analysis

Yasunori Okabe

1. Introduction

With Einstein's theory of Brownian motion [6] and Langevin's theory of stochastic differential equations [16] in statistical physics as its origin, the present theory of diffusion processes in probability theory teaches us the following: the infintesimal generators of semigroups associated with the diffusion processes and their time evolution can be characterized by the qualitative nature of diffusion properties derived from Kolmogorov's second-order differential equations of elliptic type and from Ito's stochastic differential equations of Markovian type, respectively.

In the paper "On a Langevin equation" which appeared ten years ago in [24], we introduced the property of T-positivity, which is broader than the property of diffusion, from the requirements in quantum field theory and showed that this qualitative property determines the $[\alpha, \beta, \gamma]$-Langevin equation ($\equiv$ model) in the latter sense stated above. This study was started with the aim of clarifying the mathematical structure of the fluctuation-dissipation theorem, which is one of the fundamental principles in nonequilibrium statistical physics.

An investigation that cannot but throw doubt on Einstein's theory, however, has already been presented in the 1960's. In the study in statistical physics by Alder and Wainwright [2], a computer simulation of a hydrodynamic model compounded of a hard sphere and a hard disk, it was reported that the motion of the hard sphere has the so-called Alder-Wainwright effect; that is, the velocity correlation function decreases not exponentially, but polynomially.

The study of such an Alder-Wainwright effect, from the viewpoint of statistical physics, was carried out by applying Kubo's linear response theory [13] to the study by Stokes and Boussinesq [41, 3] in the late nineteenth century before the one by Einstein and Langevin. The Alder-Wainwright effect is now verified in both theory and experiment [43, 15, 40]. In the course of these researches, it has been recognized that the model for the velocity of Brownian motion with the Alder-Wainwright effect can be described as a stationary solution of the Stokes-Boussinesq-Langevin equation whose approximate accuracy is thermodynamically better than the classical Langevin equation, that is, Ornstein-Uhlenbeck's Langevin equation.

We have perceived that this stationary solution has T-positivity from the point of view of the theory of stochastic processes, and we have felt keenly that it is

---

1991 *Mathematics Subject Classification.* Primary 60H15.

This article originally appeared in Japanese in Sûgaku **43** (4) (1991), 322–346.

necessary to remove the conditions posed in [20]–[24] as a requirement from statistical physics. This has propelled our subsequent research. In both the continuous case and the discrete case, we have obtained KMO-Langevin equations describing the time evolution of weakly stationary processes with T-positivity as a general form of the Stokes-Boussinesq-Langevin equation and clarified the mathematical structure of the Alder-Wainwright effect [27]–[34]. Thus, the $[\alpha, \beta, \gamma]$-Langevin equations in [22] changed into KMO-Langevin equations that have sufficient significance from the viewpoint of both probability theory and statistical physics. Furthermore, in the discrete case, with an aim of obtaining a certain algorithm available to computer science in technology, social science, and so on, KMO-Langevin equations, with infinite delay drift terms, offered their field of research to KM_2O-Langevin equations with finite delay drift terms that characterize local and weak stationarity of stochastic processes with discrete time [35, 39, 38]. The guiding principle is the following philosophy:

FDP (Fluctuation-Dissipation Principle). In order to study natural and social phenomena with random time evolution, we should not do an a priori argument in model building. At first we observe certain qualitative properties (e.g., local and weak stationarity) to characterize stochastic models to which we want to apply the theory of stochastic processes, and then we check its property from given data. It is important for us to take out a random force giving rise to its complexity from the data, according to the fluctuation-dissipation theorem.

Then **FDP** becomes really effective in causal analysis. If a certain causal relation between two kinds of data that vary randomly according to the change of time were asserted on the basis of certain models constructed deductively, then we would not be able to have any objective persuasive power. Thus, we propose a stationary test—Test(S)—checking the local and weak stationarity of time series formed by a given data and a causal test—Test(CS)—judging the existence and direction of a causal relation between two kinds of given data: At first we check local and weak stationarity for stochastic process with a given data as its realization on the basis of the theory of KM_2O-Langevin equations (Test(S)). If the given data does not pass Test(S), then apply appropriate injective transformations to the original data and then examine Test(S) for the new data. We can proceed to Test(CS) only after passing Test(S), because our Test(CS) is effective only for data passing Test(S).

However, the causal relation among things becomes a serious problem when certain abnormal phenomena occur. For example, certain abnormal phenomena such as the warming of the climate, earthquakes, the eruption of volcanos, and so on, in the environment and certain financial panics, such as a slump in shares in the economic world, will destroy the local and weak stationarity for time series. Thus, we need some effective transformations which reproduce the local and weak stationarity. For that purpose, we consider the following three kinds of transformations: (i) difference; (ii) logarithm (for positive data); (iii) arctangent. The important fact is that if these transformed data pass both Test(S) and Test(CS), then we can objectively assert the existence and direction of a causal relation among the given original data, taking into account particularly that (ii) and (iii) have inverse transformations.

In this paper, we shall present (dissipate) our studies (fluctuation) during the ten years since the paper [24]. We would like to seek constructive criticism from both pure mathematicians and applied mathematicians to make our guiding principle **FDP** more and more powerful.

2. Stokes-Boussinesq-Langevin equation

We shall consider a model which manifests the motion of a hard sphere of radius r and mass m moving with velocity $X(t)$ at time t in a fluid with viscosity η and density ρ, subject to a random force $W(t)$ and a drag force $F(t)$ at time t. Since, in this case, Newton's equation becomes

$$(2.1) \qquad m\dot{X}(t) = -F(t) + W(t) \quad \text{in } \mathbb{R},$$

we can take the inverse Fourier transform of the both sides of equation (2.1) to find

$$(2.2) \qquad m(-i\xi)\widetilde{X}(\xi) = -\widetilde{F}(\xi) + \widetilde{W}(\xi) \quad \text{in } \mathbb{R}.$$

This, for almost all $\xi \in \mathbb{R}$, can be rewritten as

$$(2.3) \qquad m\frac{d[e^{-it\xi}\widetilde{X}(\xi)]}{dt} = -e^{-it\xi}\widetilde{F}(\xi) + e^{-it\xi}\widetilde{W}(\xi) \quad \text{in } \mathbb{R}.$$

Equation (2.3) represents the model for the hard sphere of radius r and mass m vibrating with frequency ξ and velocity $e^{-it\xi}\widetilde{X}(\xi)$, subject to a fluctuating force $e^{-it\xi}\widetilde{W}(\xi)$ and a drag force $e^{-it\xi}\widetilde{F}(\xi)$ at time t.

By solving a linearized Navier-Stokes equation subject to incompressibility and stick boundary conditions in hydrodynamics, Stokes [42] showed that the drag force in equation (2.3) is given by

$$e^{-it\xi}\widetilde{F}(\xi) = 6\pi r\eta \left\{1 + r\left(\frac{\xi\rho}{2\eta}\right)^{1/2}\right\} e^{-it\xi}\widetilde{X}(\xi)$$
$$+ 3\pi r^2 \left(\frac{2\rho\eta}{\xi}\right)^{1/2} \left\{1 + \frac{2r}{9}\left(\frac{\xi\rho}{2\eta}\right)^{1/2}\right\} \frac{d\{e^{-it\xi}\widetilde{X}(\xi)\}}{dt}$$

and hence

$$(2.4) \quad \widetilde{F}(\xi) = \left\{6\pi r\eta + 6\pi r^2\sqrt{\frac{\rho\eta\xi}{2}} + \left(3\pi r^2\sqrt{\frac{2\rho\eta}{\xi}} + \frac{2\pi r^3}{3}\rho\right)(-i\xi)\right\}\widetilde{X}(\xi).$$

Moreover, Boussinesq [3] took the Fourier transform of the both sides of equation (2.4) to obtain the drag force $F(t)$ in equation (2.1):

$$(2.5) \quad F(t) = 2\pi r^3\rho\left\{\frac{1}{3}\frac{dX(t)}{dt} + \frac{3\eta}{r^2\rho}X(t) + \frac{3}{r}\left(\frac{\eta}{\pi\rho}\right)^{1/2}\int_{-\infty}^{t}\frac{1}{\sqrt{t-s}}\frac{dX(s)}{ds}ds\right\}.$$

Hence, equation (2.1) becomes

$$(2.6) \quad m^*\dot{X}(t) = -6\pi r\eta X(t) - 6\pi r^2\left(\frac{\rho\eta}{\pi}\right)^{1/2}\int_{-\infty}^{t}\frac{1}{\sqrt{t-s}}\dot{X}(s)\,ds + W(t),$$

where m^* is the effective mass given by

$$(2.7) \qquad m^* = m + \frac{2}{3}\pi r^3 \rho.$$

This equation (2.6) is nothing but the Stokes-Boussinesq-Langevin equation [44, 15, 28]. From the point of view of the theory of stochastic processes, we shall consider mathematical justification for the singular integral term on the right-hand side of equation (2.6) and a mathematical meaning for its solution $X(t)$.

Substituting (2.4) into (2.2), we obtain

$$\tilde{X}(\xi) = \{\sqrt{2\pi} h_{SB}(\xi)\} \widetilde{W}(\xi), \tag{2.8}$$

which is a relation representing equation (2.6) in the frequency world. Here h_{SB} is the frequency response function on $(\mathbb{R} - \{0\}) \cup \mathbb{C}^+$ defined by

$$h_{SB}(\zeta) = \frac{1}{\sqrt{2\pi}} \frac{1}{6\pi r \eta + m^*(-i\zeta) + 6\pi r^2 \sqrt{\rho\eta}\sqrt{-i\zeta}}. \tag{2.9}$$

Furthermore, $\sqrt{-i\zeta}$ stands for $\sqrt{-i\zeta} = \exp\frac{1}{2}(\log|\zeta| + i\,\mathrm{Arg}(-i\zeta))$. By using the formula

$$\int_0^\infty \frac{1}{\lambda - i\zeta} \frac{1}{\sqrt{\lambda}} d\lambda = \pi \frac{1}{\sqrt{-i\zeta}} \qquad (\zeta \in \mathbb{C}^+), \tag{2.10}$$

we can rewrite the function h_{SB} as

$$h_{SB}(\zeta) = \frac{\alpha_{SB}}{\sqrt{2\pi}} \frac{1}{\beta_{SB} - i\zeta - i\zeta \int_0^\infty (\lambda - i\zeta)^{-1} \rho_{SB}(d\lambda)}, \tag{2.11}$$

where α_{SB} and β_{SB} are positive constants and ρ_{SB} is a Borel measure on $[0, \infty)$ given by

$$(\alpha_{SB}, \beta_{SB}, \rho_{SB}) = \left(\frac{1}{m^*}, \frac{6\pi r\eta}{m^*}, \frac{6r^2\sqrt{\rho\eta}}{m^*} \frac{1}{\sqrt{\lambda}} d\lambda\right). \tag{2.12}$$

It can be shown that h_{SB} becomes an outer function as an L^2-function and its Fourier transform $\hat{h}_{SB}$ has the following form:

$$\hat{h}_{SB}(t) = \chi_{[0,\infty)}(t) \int_0^\infty e^{-t\lambda} v_{SB}(d\lambda), \tag{2.13}$$

$$v_{SB}(d\lambda) = \left(\frac{2}{\pi}\right)^{1/2} \alpha_{SB} e_{SB} \frac{\sqrt{\lambda}}{(\beta_{SB} - \lambda)^2 + e_{SB}^2 \lambda} d\lambda. \tag{2.14}$$

Here the positive constant e_{SB} is defined by

$$e_{SB} = 6\pi r^2 \frac{\sqrt{\rho\eta}}{m^*}. \tag{2.15}$$

Concerning the fluctuating force $\mathbf{W} = (W(t); t \in \mathbb{R})$, at first we treat white noise, which is easy to understand mathematically:

$$W(t) = \alpha_W \dot{B}(t), \tag{2.16}$$

where α_W is a positive constant and $\mathbf{B} = (B(t); t \in \mathbb{R})$ is a standard Brownian motion defined on a probability space $(\Omega, \mathscr{B}, P)$. It then follows that the stochastic process

$\mathbf{X_W} = (X_W(t); t \in \mathbb{R})$ constructed from $X(t)$ ($t \in \mathbb{R}$) that is determined by relation (2.8), can be represented as

$$(2.17) \qquad X_W(t) = \frac{\alpha_W}{\sqrt{2\pi}} \int_{\mathbb{R}} \hat{h}_{SB}(t-u) \, dB(u).$$

We then find that $\mathbf{X_W}$ has weak stationarity:

$$(2.18) \qquad \int_{\Omega} X(t) X(s) \, dP = R_W(t-s).$$

Moreover, we see from (2.13), (2.14), and (2.18) that the correlation function R_W of the process $\mathbf{X_W}$ can be calculated as follows:

$$(2.19) \qquad R_W(t) = \frac{\alpha_W^2}{2\pi} \int_{\mathbb{R}} \hat{h}_{SB}(t+u) \hat{h}_{SB}(u) \, du = \alpha_W^2 \int_0^{\infty} e^{-|t|\lambda} \sigma_{SB}(d\lambda),$$

$$(2.20) \qquad \sigma_{SB}(d\lambda) = \frac{\alpha_{SB}^2}{\pi} \frac{e_{SB}}{\beta_{SB} + \lambda + e_{SB}\sqrt{\lambda}} \frac{\sqrt{\lambda}}{(\beta_{SB} - \lambda)^2 + e_{SB}^2 \lambda} \, d\lambda.$$

We shall now consider the case where the fluctuating force $\mathbf{W} = (W(t); t \in \mathbb{R})$ is Kubo noise $\mathbf{I} = (I(\varphi); \varphi \in \mathcal{S}(\mathbb{R}))$. It follows from Kubo's linear response theory [13, 14] that the random force $\mathbf{I}$ to be defined and the weakly stationary process $\mathbf{X_K} = (X_K(t); t \in \mathbb{R})$, which is a solution of equation (2.6) when W is replaced by I, are characterized by the following relation:

$$(2.21) \qquad \int_{\mathbb{R}} X_K(u) \varphi(u) \, du = \frac{1}{\sqrt{2\pi}} \int_0^{\infty} R_K(u) I(\varphi(\cdot + u)) \, du \qquad (\varphi \in \mathcal{S}(\mathbb{R})).$$

Here R_K is the correlation function of $\mathbf{X_K}$:

$$(2.22) \qquad R_K(t-s) \equiv \int_{\Omega} X_K(t)(\omega) X_K(s)(\omega) \, dP(\omega).$$

By applying the inverse Fourier transform to the both sides of equation (2.21), we see that it is necessary for the following relation to hold, in order to arrive at (2.8) with W replaced by I,

$$(2.23) \qquad R_K(t) = \hat{h}_{SB}(|t|) = \int_0^{\infty} e^{-|t|\lambda} \nu_{SB}(d\lambda).$$

We shall use the inverse argument. Since the function on the right-hand side of (2.23) is always nonnegative definite, there exists a weakly stationary process $\mathbf{X_K} = (X_K(t); t \in \mathbb{R})$ with the function R_K as its correlation function. Further, we find that the stochastic process determined by relation (2.8) is nothing but the stationary random distribution $\mathbf{I} = (I(\varphi); \varphi \in \mathcal{S}(\mathbb{R}))$ to be called *Kubo noise*.

This is the prescription to let us construct the stationary solution $\mathbf{X_K}$ and Kubo noise $\mathbf{I}$ from our equation (2.6) ($\equiv$ model), which is the essential part of Kubo's linear response theory. We used a good deal of labor to establish this prescription [27, 28]. In reality, the research to deduce the mathematical structure of Kubo's fluctuation-dissipation theorem in statistical physics depended entirely upon the introduction of Kubo noise, because there was no notion of Kubo noise in Kubo's linear response theory.

We might treat other noises besides white noise and Kubo noise as random forces in equation (2.6). The approach we should take according to the **FDP** stated in §1 is to answer the following questions:

(∗) $\begin{cases} \text{What kind of qualitative nature does the solution } (\mathbf{X_W}, \mathbf{X_K}) \\ \text{of equation (2.6) have?} \end{cases}$

(∗∗) $\begin{cases} \text{What kinds of equations can govern a stochastic process} \\ \text{with such a qualitative nature?} \end{cases}$

By (2.19) and (2.23), we can give the question (∗) the answer that both $\mathbf{X_W}$ and $\mathbf{X_K}$ have *T-positivity*. We discuss its definition and question (∗∗) in what follows.

3. *T*-positivity and KMO-Langevin equation (continuous system)

Let $\mathbf{X} = (X(t); t \in \mathbb{R})$ be a real weakly stationary process with expectation 0 whose correlation function R has the following representation called *T-positivity*:

$$(3.1) \qquad R(t) = \int_{[0,\infty)} e^{-|t|\lambda} \sigma(d\lambda) \qquad (t \in \mathbb{R}).$$

Here we put the integrability condition on the Borel measure σ on $[0, \infty)$:

$$(3.2) \qquad \sigma(\{0\}) = 0, \quad 0 < \sigma([0,\infty)) < \infty,$$

$$(3.3) \qquad \int_0^\infty \lambda^{-1} \sigma(d\lambda) < \infty,$$

$$(3.4) \qquad \int_0^\infty \lambda \sigma(d\lambda) < \infty.$$

We denote by $\Delta = \Delta(\xi)$ ($\xi \in \mathbb{R}$) the spectral density of the correlation function R:

$$(3.5) \qquad R(t) = \int_\mathbb{R} e^{-it\xi} \Delta(\xi) \, d\xi \qquad (t \in \mathbb{R}).$$

It then follows from (3.1) that

$$(3.6) \qquad \Delta(\xi) = \frac{1}{\pi} \int_0^\infty \frac{\lambda}{\lambda^2 + \xi^2} \sigma(d\lambda) \qquad (\xi \in \mathbb{R}).$$

Moreover, let h and $[R]$ be the outer function and the complex mobility function of $\mathbf{X}$, respectively:

$$(3.7) \qquad h(\zeta) = \exp\left(\frac{1}{2\pi i} \int_\mathbb{R} \frac{1 + \lambda \zeta}{\lambda - \zeta} \frac{\log \Delta(\lambda)}{1 + \lambda^2} \, d\lambda\right) \qquad (\zeta \in \mathbb{C}^+),$$

$$(3.8) \qquad [R](\zeta) = \frac{1}{2\pi} \int_0^\infty e^{i\zeta t} R(t) \, dt \qquad (\zeta \in \mathbb{C}^+).$$

We approximate the Borel measure σ by a sequence $(\sigma_n; n \in \mathbb{N})$ of Borel measures satisfying, together with (3.2) and (3.3), the condition

$$(3.9) \qquad \int_0^\infty \lambda^2 \sigma(d\lambda) < \infty,$$

which is stronger than (3.4). This was treated in **[21]–[24]**. Applying the results in

[21] we obtain the representation theorem for the outer function h and the complex mobility function $[R]$.

THEOREM 3.1 ([27]). *There exist uniquely two triplets $(\alpha_j, \beta_j, \rho_j)$ $(j = 1, 2)$ satisfying the following properties*:
  (i) $\alpha_j > 0$, $\beta_j > 0$,
  (ii) ρ_j *is a Borel measure on* $[0, \infty)$ *with* $\rho_j(\{0\}) = 0$ *and* $\int_0^\infty \frac{1}{\lambda+1} \rho_j(d\lambda) < \infty$,
  (iii) $h(\zeta) = \frac{\alpha_1}{\sqrt{2\pi}} \frac{1}{\beta_1 - i\zeta - i\zeta \int_0^\infty (\lambda - i\zeta)^{-1} \rho_1(d\lambda)}$ $(\zeta \in \mathbb{C}^+)$,
  (iv) $[R](\zeta) = \frac{\alpha_2}{\sqrt{2\pi}} \frac{1}{\beta_2 - i\zeta - i\zeta \int_0^\infty (\lambda - i\zeta)^{-1} \rho_2(d\lambda)}$ $(\zeta \in \mathbb{C}^+)$.

DEFINITION 3.1. We shall call the triplet $(\alpha_1, \beta, \rho_1)$ (resp., $(\alpha_2, \beta_2, \rho_2)$) the *first* (resp., *the second*) *KMO-Langevin data associated with the correlation function R.*

REMARK 3.1. The integrability condition (3.4) is not necessary for the introduction of the second KMO-Langevin data $(\alpha_2, \beta_2, \rho_2)$.

For the purpose of substantiating this nomenclature, we shall derive a stochastic differential equation describing the time evolution of **X**, which will simultaneously answer the problem $(**)$ stated in the last part of §2.

From Karhunen's representation theorem [5] in the spectral theory for the weakly stationary process **X** we find that there exists an orthogonal process $\mathbf{B} = (B(t); t \in \mathbb{R})$ satisfying the following: for any $s, t \in \mathbb{R}$,

$$(3.10) \qquad \int_\Omega B(s) B(t) \, dP = s \wedge t,$$

$$(3.11) \qquad X(t) = \frac{1}{\sqrt{2\pi}} \int_{-\infty}^t \hat{h}(t - u) \, dB(u),$$

(3.12)
the linear hull of $\{X(u); u \leq t\}$ = the linear hull of $\{B(u) - B(v); u, v \leq t\}$.

The random stationary distribution $\dot{\mathbf{B}}$ resulting from differentiation of **B** is called the standard white noise. On the other hand, Kubo noise $\mathbf{I} = (I(\varphi); \varphi \in \mathcal{S}(\mathbb{R}))$ is the random stationary distribution defined by

$$(3.13) \qquad I(\varphi) = \int_\mathbb{R} \left(\frac{h}{[R]} \hat{\varphi} \right)^\sim (s) \, dB(s).$$

Corresponding to (3.11) and (3.12), we have the following: for any $t \in \mathbb{R}$, $\varphi \in \mathcal{S}(\mathbb{R})$,

$$(3.14) \qquad \int_\mathbb{R} X(u) \varphi(u) \, du = \frac{1}{\sqrt{2\pi}} \int_0^\infty R(u) I(\varphi(\cdot + u)) \, du,$$

(3.15)
the linear hull of $\{X(u); u \leq t\}$ = the linear hull of $\{I(\varphi); \operatorname{supp} \varphi \subset (-\infty, t]\}$.

We set $\varphi = \delta(\cdot - t)$ to obtain an intuitive representation for (3.13) and (3.14). We then find that (3.13) can be rewritten as

$$(3.13)' \qquad I(t) = \frac{1}{2\pi} \int_{-\infty}^t \left(\frac{h}{[R]} \right)^\wedge (t - u) \, dB(u)$$

and (3.14) has the following representation analogous to (3.11):

$$(3.14)' \qquad X(t) = \frac{1}{\sqrt{2\pi}} \int_{-\infty}^{t} R(t-u)I(u)\,du.$$

We have derived the standard white noise $\dot{\mathbf{B}}$ and Kubo noise $\mathbf{I}$ as random forces associated with the given weakly stationary process $\mathbf{X}$. The representation (3.11) (resp., (3.14)′) can be regarded as a construction of a model such that $\mathbf{X}$ is the common output, $\dot{\mathbf{B}}$ (resp., $\mathbf{I}$) is the input, and $\hat{h}$ (resp., $\chi_{[0\infty,)}R$) is the response function. Following the idea in **FDP**, we find that these models can be rewritten into the following equations.

THEOREM 3.2 ([27]). *As random distributions, $\mathbf{X}$ and $\dot{\mathbf{B}}$ (resp., $\mathbf{X}$ and $\mathbf{I}$) satisfy the following stochastic differential equation (3.16) (resp., (3.17)):*

$$(3.16) \qquad \dot{\mathbf{X}} = -\beta_1 \mathbf{X} - \lim_{\varepsilon \downarrow 0} \gamma_{1,\varepsilon} * \dot{\mathbf{X}} + \alpha_1 \dot{\mathbf{B}},$$

$$(3.17) \qquad \dot{\mathbf{X}} = -\beta_2 \mathbf{X} - \lim_{\varepsilon \downarrow 0} \gamma_{2,\varepsilon} * \dot{\mathbf{X}} + \alpha_2 \mathbf{I}.$$

Here $\gamma_{j,\varepsilon}$ $(j = 1, 2)$ are the functions given by

$$(3.18) \qquad \gamma_{j,\varepsilon}(t) = \chi_{(0,\infty)}(t) \int_{\varepsilon}^{\infty} e^{-t\lambda} \rho_j(d\lambda).$$

DEFINITION 3.2. We shall call the stochastic differential equation (3.16) (resp., (3.17)) the first (resp., the second) KMO-Langevin equation associated with the weakly stationary process $\mathbf{X}$.

EXAMPLE 3.1. For each $v > 0$, $\beta > 0$, let $\mathbf{X}_{v,\beta}$ be a weakly stationary process with correlation function $R_{v,\beta}$ of the form

$$(3.19) \qquad R_{v,\beta}(t) = v e^{-\beta |t|} \qquad (t \in \mathbb{R}).$$

This is the case where σ in (3.1) is a point measure $\sigma = v\delta_\beta$. The process $\mathbf{X}_{v,\beta}$ is called Ornstein-Uhlenbeck's Brownian motion and was investigated in a study by Einstein and Langevin. It follows that the outer function $h_{v,\beta}$ and the complex mobility function $[R_{v,\beta}]$ are given by the following:

$$(3.20) \qquad h_{v,\beta}(\zeta) = \frac{\sqrt{2v\beta}}{2\pi} \frac{1}{\beta - i\zeta} \quad \text{and} \quad [R_{v,\beta}](\zeta) = \frac{v}{2\pi} \frac{1}{\beta - i\zeta} \qquad (\zeta \in \mathbb{C}^+).$$

Hence, the first and the second KMO-Langevin data associated with the correlation $R_{v,\beta}$ become

$$(3.21) \qquad (\alpha_1, \beta_1, \rho_1) = (\sqrt{2v\beta}, \beta, 0) \quad \text{and} \quad (\alpha_2, \beta_2, \rho_2) = (v, \beta, 0).$$

Furthermore, $\mathbf{X}_{v,\beta}$ satisfy the following first and second KMO-Langevin equations:

$$(3.22) \qquad \dot{\mathbf{X}}_{v,\beta} = -\beta \mathbf{X}_{v,\beta} + \sqrt{2v\beta}\, \dot{\mathbf{B}}_{v,\beta},$$

$$(3.23) \qquad \dot{\mathbf{X}}_{v,\beta} = -\beta \mathbf{X}_{v,\beta} + v \mathbf{I}_{v,\beta}.$$

Here $\dot{\mathbf{B}}_{v,\beta}$ and $\mathbf{I}_{v,\beta}$ are the standard white noise and Kubo noise associated with $\mathbf{X}_{v,\beta}$, respectively. It follows from (3.22) and (3.23) that both random distributions are equal up to positive constants:

$$\mathbf{I}_{v,\beta} = \sqrt{\frac{2\beta}{v}} \dot{\mathbf{B}}_{v,\beta}. \tag{3.24}$$

4. T-positivity and KMO-Langevin equation (discrete system)

Let $\mathbf{X} = (X(n); n \in \mathbb{Z})$ be a real weakly stationary process defined on a probability space $(\Omega, \mathscr{B}, P)$ with expectation 0 and correlation function R of the form (T-positivity):

$$R(n) = \int_{[-1,1]} t^{|n|} \sigma(dt) \qquad (n \in \mathbb{Z}). \tag{4.1}$$

Here the Borel measure σ on $[-1, 1]$ satisfies the following integrability conditions:

$$\sigma(\{-1, 1\}) = 0, \qquad 0 < \sigma([-1, 1]) < \infty, \tag{4.2}$$

$$\int_{-1}^{1} \left(\frac{1}{1+t} + \frac{1}{1-t} \right) \sigma(dt) < \infty. \tag{4.3}$$

The spectral measure $\Delta = \Delta(\theta)$ of the correlation function R determined by the relation

$$R(n) = \int_{-\pi}^{\pi} e^{-in\theta} \Delta(\theta) \, d\theta \qquad (n \in \mathbb{Z}) \tag{4.4}$$

satisfies

$$\Delta(\theta) = \frac{1}{\pi} \int_{-1}^{1} \frac{1 - t^2}{|1 - te^{i\theta}|^2} \sigma(dt) \qquad (\theta \in (-\pi, \pi)). \tag{4.5}$$

The outer function h and the complex mobility function $[R]$ of the process $\mathbf{X}$ are respectively defined by

$$h(z) = \exp\left(\frac{1}{4\pi} \int_{-\pi}^{\pi} \frac{e^{i\theta} + z}{e^{i\theta} - z} \log \Delta(\theta) \, d\theta \right) \qquad (z \in \mathbb{C}, |z| < 1), \tag{4.6}$$

$$[R](z) = \frac{1}{2\pi} \sum_{n=0}^{\infty} R(n) z^n \qquad (z \in \mathbb{C}, |z| < 1). \tag{4.7}$$

By using a homeomorphism $\varphi: \varphi(t) = (1 + t)/(t - 1)$ from $(-1, 1)$ in $[0, \infty)$, we transform the Borel measure σ on $(-1, 1)$ to the Borel measure $\sigma_c = \varphi(\sigma)$ on $[0, \infty)$ and apply Theorem 3.1 for the continuous case. We obtain the representation theorem for the outer function h and the complex mobility function $[R]$ in the discrete case:

THEOREM 4.1 ([32, 33]). *There exist uniquely two triplets* $(\alpha_j, \beta_j, \rho_j)$ $(j = 1, 2)$ *satisfying the following properties*:
  (i) $\alpha_j > 0$, $\beta_j > 0$,
  (ii) ρ_j *is a Borel measure on* $[-1, 1]$ *with* $\rho_j(\{-1, 1\}) = 0$ *and* $\rho_j([-1, 1]) < \infty$,
  (iii) $h(z) = \frac{\alpha_1}{\sqrt{2\pi}} \frac{1}{\beta_1(1+z) + 1 - z + (1-z^2) \int_{-1}^{1} (1-tz)^{-1} \rho_1(dt)}$ $(z \in \mathbb{C}, |z| < 1)$,

(iv) $[R](z) = \frac{\alpha_2}{\sqrt{2\pi}} \frac{1}{\beta_2(1+z)+1-z+(1-z^2)\int_{-1}^{1}(1-tz)^{-1}\rho_2(dt)}$ $(z \in \mathbb{C}, |z| < 1)$.

DEFINITION 4.1. We shall call the triplet $(\alpha_1, \beta_1, \rho_1)$ (resp., $(\alpha_2, \beta_2, \rho_2)$) the first (resp., the second) KMO-Langevin data associated with the correlation function R.

We shall derive a stochastic difference equation describing the time evolution of **X** in the discrete system, which will answer the discrete version of the question $(**)$ in §2. It follows from the spectral theory of the weakly stationary process **X** that there exists a weakly stationary process $\xi = (\xi(n); n \in \mathbb{Z})$, to be called the standard white noise, satisfying the following:

$$\text{(4.8)} \quad \int_\Omega \xi(n)\xi(m)\,dP = \delta_{nm} \quad (n, m \in \mathbb{Z}),$$

$$\text{(4.9)} \quad X(n) = \frac{1}{\sqrt{2\pi}} \sum_{m=-\infty}^{n} \hat{h}(n-m)\xi(m) \quad (n \in \mathbb{Z}),$$

$$\text{(4.10)} \quad \mathbf{M}^n_{-\infty}(\mathbf{X}) = \mathbf{M}^n_{-\infty}(\xi) \quad (n \in \mathbb{Z}).$$

In general, for an $\mathbb{R}^d$-valued stochastic process $\mathbf{Y} = ({}^t(Y_1(n),\ldots,Y_d(n)); n \in \mathbb{Z})$ defined on the probability space $(\Omega, \mathcal{B}, P)$, we shall denote by $\mathbf{M}_n^m(\mathbf{Y})$ the closed subspace of $L^2(\Omega, \mathcal{B}, P)$ for $n, m, -\infty \leq n \leq m \leq \infty$:

$$\text{(4.11)} \quad \mathbf{M}_n^m(Y) \equiv \text{the linear hull of } \{Y_j(k); 1 \leq j \leq d,\ n \leq k \leq m\}.$$

Corresponding to the intuitive representation $(3.13)'$ in the continuous system, Kubo noise $\mathbf{I} = (I(n); n \in \mathbb{Z})$ in the discrete system is a weakly stationary process defined by

$$\text{(4.12)} \quad I(n) = \frac{1}{2\pi} \sum_{m=-\infty}^{n} \widehat{\left(\frac{h}{[R]}\right)}(n-m)\xi(m).$$

As the discrete version of (3.14) ((3.14)$'$) and (3.15) in the continuous system, we have

$$\text{(4.13)} \quad X(n) = \frac{1}{\sqrt{2\pi}} \sum_{m=-\infty}^{n} R(n-m)I(m) \quad (n \in \mathbb{Z}),$$

$$\text{(4.14)} \quad \mathbf{M}^n_{-\infty}(\mathbf{X}) = \mathbf{M}^n_{-\infty}(\mathbf{I}) \quad (n \in \mathbb{Z}).$$

Similarly, as in the continuous system, the standard white noise ξ and the Kubo noise **I** were derived starting from the given weakly stationary process **X**. The model (4.9) (resp., (4.13)) in which **X** is the common output, ξ (resp., **I**) is the input, and $\hat{h}$ (resp., $\chi_{[0,\infty)}R$) is the response function can be described by the following stochastic difference equation (4.15) (resp., (4.16)):

THEOREM 4.2 ([**32, 33**]). *The process* **X** *satisfies*

(4.15) $\quad X(n) - X(n-1) = -\beta_1(X(n) + X(n-1)) - (\gamma_1 * X)(n) + \alpha_1 \xi(n);$

(4.16) $\quad X(n) - X(n-1) = -\beta_2(X(n) + X(n-1)) - (\gamma_2 * X)(n) + \alpha_2(n).$

Here γ_j $(j = 1, 2)$ *are* $l^1(\mathbb{Z})$-*functions given by*

$$\text{(4.17)} \quad \gamma_j = \frac{1}{2\pi}\widehat{\left((1-e^{2i})\int_{-1}^{1}\frac{1}{1-te^{i\cdot}}\rho_j(dt)\right)}.$$

DEFINITION 4.2. We shall call the stochastic difference equation (4.15) (resp., (4.16)) the first (resp., the second) KMO-Langevin equation associated with the weakly stationary process **X**.

The functions γ_j ($j = 1, 2$) determined by (4.17) can be represented by

$$(4.18) \quad \gamma_j(n) = \begin{cases} 0 & (n = -1, -2, \ldots), \\ \int_{-1}^{1} t^n \rho_j(dt) & (n = 0, 1), \\ \int_{-1}^{1} (t^n - t^{n-2}) \rho_j(dt) & (n = 2, 3, \ldots), \end{cases}$$

and hence

$$(4.19) \quad \sum_{n=0}^{\infty} \gamma_j(n) = 0 \quad \text{and} \quad \sum_{n=0}^{\infty} (-1)^n \gamma_j(n) = 0.$$

Inoue [11] noticed the following fact: the first (resp., the second) KMO-Langevin equation (4.15) (resp., (4.16)) can be rewritten in the same form as the delay term in the first (resp., the second) KMO-Langevin equation (3.16) (resp., (3.17)). Defining the functions $\gamma_{j,0}$ ($j = 1, 2$) by

$$(4.20) \quad \gamma_{j,0}(n) = \begin{cases} 0 & (n = -1, -2, \ldots), \\ \rho_j([-1, 1]) & (n = 0), \\ \int_{-1}^{1} (t^n + t^{n-1}) \rho_j(dt) & (n = 1, 2, \ldots), \end{cases}$$

we have

$$(4.21) \quad \gamma_{j,0}(n) - \gamma_{j,0}(n-1) = \gamma_j(n) \quad (n \in \mathbb{Z}).$$

This can allow us to rewrite the delay term $(\gamma_j * X)(n)$ in (4.15) and (4.16) as

$$(4.22) \quad (\gamma_j * X)(n) = \underset{\varepsilon \downarrow 0}{\text{l.i.m.}} \sum_{m=-\infty}^{n-1} \gamma_{j,0}^\varepsilon(n-m) \Delta X(m) + \gamma_j(0) \Delta X(n).$$

Here we put

$$(4.23) \quad \Delta X(n) = X(n) - X(n-1),$$

$$(4.24) \quad \gamma_{j,0}^\varepsilon(n) = \begin{cases} 0 & (n = -1, -2, \ldots), \\ \rho_j([-1+\varepsilon, 1-\varepsilon]) & (n = 0), \\ \int_{-1+\varepsilon}^{1-\varepsilon} (t^n + t^{n-1}) \rho_j(dt) & (n = 1, 2, \ldots). \end{cases}$$

Therefore, for each $j = 1, 2$, we can set

$$(4.25) \quad (\alpha_{m,j}, \beta_{m,j}, \gamma_{m,j}) = \left(\frac{\alpha_j}{1 + \gamma_j(0)}, \frac{2\beta_j}{1 + \gamma_j(0)}, \frac{\gamma_{j,0}}{1 + \gamma_j(0)} \right)$$

in order to rewrite equation (4.15) (resp., (4.16)) as equation (4.26) (resp., (4.27)):

$$(4.26) \quad \begin{aligned} \Delta X(n) = & -\beta_{m,1} \frac{(X(n) + X(n-1))}{2} \\ & - \underset{\varepsilon \downarrow \varepsilon}{\text{l.i.m.}} \sum_{m=-\infty}^{n-1} \gamma_{1,0}^\varepsilon(n-m) \Delta X(m) + \alpha_{m,1} \xi(n), \end{aligned}$$

$$\Delta X(n) = -\beta_{m,2}\frac{(X(n)+X(n-1))}{2}$$
(4.27)
$$-\operatorname*{l.i.m.}_{\varepsilon\downarrow 0}\sum_{m=-\infty}^{n-1}\gamma_{2,0}^{\varepsilon}(n-m)\Delta X(m) + \alpha_{m,2}I(n).$$

DEFINITION 4.3. We shall call the stochastic differential equation (4.26) (resp., (4.27)) the modified first (resp., the modified second) KMO-Langevin equation associated with the weakly stationary process **X**. The triple $(\alpha_{m,1},\beta_{m,1},\gamma_{m,1})$ (resp., $(\alpha_{m,2},\beta_{m,2},\gamma_{m,2})$) is called the modified first (resp., the modified second) KMO-Langevin data associated with the correlation function R.

EXAMPLE 4.1. For each $v > 0$, $p \in (-1, 1)$, let $\mathbf{X}_{v,p}$ be a weakly stationary process with the correlation function $R_{v,p}$ of the form

(4.28)
$$R_{v,p}(n) = v p^{|n|} \qquad (n \in \mathbb{Z}).$$

This is the case where σ in (4.1) is a point measure $\sigma = v\delta_p$. The process $\mathbf{X}_{v,p}$ has the Markovian property. The outer function $h_{v,p}$ and the complex mobility function $[R_{v,p}]$ can be calculated as follows:

(4.29) $\quad h_{v,p}(z) = \sqrt{\dfrac{v}{2\pi}\dfrac{1-p^2}{1-pz}}$ and $[R_{v,p}](z) = \dfrac{v}{2\pi}\dfrac{1}{1-pz} \qquad (z \in \mathbb{C}, |z| < 1).$

Hence, the first and second KMO-Langevin data associated with the correlation $R_{v,p}$ become

(4.30)
$$\begin{cases} (\alpha_{v,p}^1, \beta_{v,p}^1, \rho_{v,p}^1) = \left(2\sqrt{v\dfrac{1-p}{1+p}}, \dfrac{1-p}{1+p}, 0\right); \\ (\alpha_{v,p}^2, \beta_{v,p}^2, \rho_{v,p}^2) = \left(\sqrt{\dfrac{2}{\pi}\dfrac{v}{1+p}}, \dfrac{1-p}{1+p}, 0\right). \end{cases}$$

Furthermore, there exists the following relation between the standard white noise $\xi_{v,p} = (\xi_{v,p}(n); n \in \mathbb{Z})$ and Kubo noise $\mathbf{I}_{v,p} = (I_{v,p}(n); n \in \mathbb{Z})$ associated with $\mathbf{X}_{v,p}$:

(4.31)
$$I_{v,p}(n) = \sqrt{2\pi(1-p)(1+p)}\,\xi_{v,p}(n).$$

5. Alder-Wainwright effect

5.1. Let $\mathbf{X}_W$ (resp., $\mathbf{X}_K$) be the stationary solution of the Stokes-Boussinesq-Langevin equation (2.6) in which the random force W is the white noise $\alpha_W \dot{\mathbf{B}}$ (resp., Kubo noise $\mathbf{I}$). It follows from the representation (2.19) and (2.23) of their correlation functions R_W and R_K that both the process $\mathbf{X}_W$ and the process $\mathbf{X}_K$ have T-positivity. Furthermore, we have the following Alder-Wainwright effect [43, 15, 28]:

ALDER-WAINWRIGHT EFFECT.

(5.1)
$$\lim_{t\to\infty}(\beta_{SB}t)^{3/2}R_W(t)$$
$$= \frac{\sqrt{\pi}R_W(0)}{2}\left\{\int_0^\infty \frac{\sqrt{y}}{(1+y+a\sqrt{y})((1-y)^2+a^2 y)}\,dy\right\}^{-1},$$

(5.2)
$$\lim_{t\to\infty}(\beta_{SB}t)^{3/2}R_K(t) = \frac{R_K(0)}{2\sqrt{\pi}}a.$$

Here it is known [15] that the constant

(5.3) $$a = \left(\frac{6\pi\gamma\rho^3}{m^*}\right)^{1/2}$$

has the following physical meaning:

(5.4) $$a = 3\left(1 + 2\frac{\rho_0}{\rho}\right)^{-1/2},$$

where ρ_0 and ρ are, respectively, the density of the hard sphere and the liquid in the same situation as in the derivation of equation (2.6).

5.2. We shall show that the mathematical background for the Alder-Wainwright effect (5.1) and (5.2) for the Stokes-Boussinesq-Langevin equation (2.6) lies in the long time behavior of the integral kernel $\sqrt{t}^{-1}$ in the delay term. Let p be a positive constant, and let L be a slowly varying function at infinity:

(5.5) $$\lim_{t \to \infty} \frac{L(st)}{L(t)} = 1 \quad \text{for any } s > 0.$$

We denote by R the correlation function of the common stationary solution $\mathbf{X}$ for both the first and the second KMO-Langevin equations (3.16) and (3.17). We then find that the Alder-Wainwright effect can be clarified as a polynomial decay for R as follows:

THEOREM 5.1 ([31, 10]). *The following are equivalent*:
 (i) $R(t) \sim t^{-(1+p)} L(t)$ as $t \to \infty$.
 (ii) $\gamma_1(t) \sim \alpha_1^{-2} \beta_1^3 p t^{-p} L(t)$ as $t \to \infty$.
 (iii) $\gamma_2(t) \sim \sqrt{2\pi}^{-1} \alpha_2^{-1} \beta_2^2 p t^{-p} L(t)$ as $t \to \infty$.

5.3. Under the condition that

(5.6) $$\rho_j([-1, 0)) = 0 \quad (j = 1, 2),$$

Inoue [11] demonstrated the Alder-Wainwright effect for both the modified first and the modified second KMO-Langevin equations (4.26) and (4.27) in the discrete system.

THEOREM 5.2 ([11]). *The following are equivalent*:
 (i) $R(n) \sim n^{-(1+p)} L(n)$ as $n \to \infty$.
 (ii) $\gamma_{m,1}(n) \sim \alpha_{m,1}^{-2} \beta_{m,1}^3 p n^{-p} L(n)$ as $n \to \infty$.
 (iii) $\gamma_{m,2}(n) \sim \sqrt{2\pi}^{-1} \alpha_{m,2}^{-1} \beta_{m,2}^2 p n^{-p} L(n)$ as $n \to \infty$.

REMARK 5.1. The condition (5.6) does not necessarily imply the following:

(5.7) $$\sigma_j([-1, 0)) = 0 \quad (j = 1, 2).$$

REMARK 5.2. We have succeeded in characterizing exponential decay of the correlation function R as that of one of the delay functions γ_j ($j = 1, 2$) in KMO-Langevin equations [37].

REMARK 5.3. The case where conditions (3.3) and (4.3) do not hold—that is, the diffusion constant in (6.1) and (6.5) of the next section is divergent—interests us in relation to certain critical phenomena of random systems and is now under investigation.

6. Fluctuation-dissipation theorem

The heart of the fluctuation-dissipation theorem (**FDT**) in **FDP** stated in §1 lies in a philosophical understanding that for any phenomena with a complex behavior, under the stationary situation, its equation of motion describing the time evolution can be separated into a random chaotic part (fluctuant term) and a dynamical calm part (dissipative term) and certain relations hold between both terms.

It seems that in the course of one's research the process of study with trouble or joy corresponds to the fluctuant part and the act of presenting the results of research corresponds to the dissipative part. We suffer criticism the more severely if the results are unexpected, but we absorb it as the fluctuating force for more research and newer presentation. The situation where we can steadily continue in such work can be called stationary.

6.1. We shall now examine how the philosophical understanding of **FDT** stated above can be mathematically represented. For the weakly stationary process **X** treated in §3, we define its diffusion constant D by

$$(6.1) \qquad D \equiv \lim_{t \to \infty} \frac{\int_\Omega (\int_0^t X(s)\,ds)^2 \, dP}{2t},$$

which can be calculated by

$$(6.2) \qquad D = \int_0^\infty R(t)\,dt = \int_0^\infty \frac{1}{\lambda}\sigma(d\lambda).$$

First we shall consider Ornstein-Uhlenbeck's Brownian motion $\mathbf{X}_{v,\beta}$ treated in Example 3.1 and explain **FDT** on the basis of the first KMO-Langevin equation (3.22). Denoting by $D_{v,\beta}$ the diffusion constant of $\mathbf{X}_{v,\beta}$, we have the following relations:

(A) $\qquad \dfrac{1}{\beta_1 - i\zeta} = \dfrac{1}{R_{v,\beta}(0)} \dfrac{h_{v,\beta}(\zeta)}{\int_\mathbb{R} \operatorname{Re} h_{v,\beta}(\xi + i0)\,d\xi};$

(B) $\qquad \dfrac{\alpha_1^2}{2} = R_{v,\beta}(0)\beta_1;$

(C) $\qquad D_{v,\beta} = \dfrac{R_{v,\beta}(0)}{\beta_1};$

(D) $\qquad D_{v,\beta} = \dfrac{\alpha_1^2}{2\beta_1^2}.$

We shall observe in turn how these relations can be extended for a general weakly stationary process **X**. Concerning relation (A), we have

A GENERALIZED **FDT** OF THE FIRST KIND ([**28**]). On the basis of the first and second KMO-Langevin equations (3.16) and (3.17), the following hold on $\mathbb{C}^+ \cup (\mathbb{R} - \{0\})$, respectively:

(A-1) $\qquad \dfrac{1}{\beta_1 - i\zeta - i\zeta \lim_{\varepsilon \downarrow 0} \int_0^\infty e^{i\zeta t}\gamma_{1,\varepsilon}(t)\,dt} = \dfrac{h(\zeta)}{\int_\mathbb{R} \operatorname{Re} h(\xi + i0)\,d\xi},$

(A-2) $\qquad \dfrac{1}{\beta_2 - i\zeta - i\zeta \lim_{\varepsilon \downarrow 0} \int_0^\infty e^{i\zeta t}\gamma_{2,\varepsilon}(t)\,dt} = \dfrac{1}{R(0)} \int_0^\infty e^{i\zeta t} R(t)\,dt.$

Concerning relation (B), we have

A GENERALIZED FDT OF THE SECOND KIND ([28]).

(B-1) $$\frac{\alpha_1^2}{2} = R(0) C_{\beta_1,\gamma_1},$$

(B-2) $$\Delta_{\alpha_2 I}(d\xi) = \frac{R(0)}{\pi} \left\{ \mathrm{Re} \left(\beta_2 - i\xi \lim_{\varepsilon \downarrow 0} \int_0^\infty e^{i\xi t} \gamma_{2,\varepsilon}(t)\, dt \right) \right\} d\xi.$$

Here C_{β_1,γ_1} is a constant defined by

(6.3) $$C_{\beta_1,\gamma_1} = \pi \left\{ \int_{\mathbb{R}} \left| \beta_1 + (-i\xi) \left(1 + \lim_{\varepsilon \downarrow 0} \int_0^\infty e^{i\xi t} \gamma_{1,\varepsilon}(t)\, dt \right) \right|^{-2} d\xi \right\}^{-1}$$

and $\Delta_{\alpha_2 I}$ stands for the spectral measure of $\alpha_2 I$.

We shall turn to Einstein's relation (C).

A GENERALIZED EINSTEIN RELATION (1) ([28]). On the basis of the first and second KMO-Langevin equations (3.16) and (3.17), we have

(C-1) $$D = \frac{R(0)}{\beta_1} \frac{C_{\beta_1,\gamma_1}}{\beta_1},$$

(C-2) $$D = \frac{R(0)}{\beta_2}.$$

Moreover, we see that

(6.4) $$\frac{C_{\beta_1,\gamma_1}}{\beta_1} \geq 1$$

and that equality holds in (6.4) only for the case where **X** is Ornstein-Uhlenbeck's Brownian motion.

Hence, a classical Einstein's relation (C) always holds on the basis of the second KMO-Langevin equation (3.17), but deviates according to the degree of the positive constant $C_{\beta_1,\gamma_1}/\beta_1 - 1$ on the basis of the first KMO-Langevin equation (3.16).

However, the situation for relation (D) is reversed.

A GENERALIZED EINSTEIN RELATION (2) ([28]). The relation

(D-1) $$D = \frac{\alpha_1^2}{2\beta_1^2}$$

always holds on the basis of the first KMO-Langevin equation (3.16). But, this does not hold on the basis of the second KMO-Langevin equation (3.17).

6.2. We shall observe in the sequel what kind of **FDT** the weakly stationary process **X** with discrete time has. **FDT** has never been established in irreversible statistical physics of a discrete system. We define a diffusion constant D associated with **X** by

(6.5)
$$D \equiv \lim_{N \to \infty} \frac{1}{2N} \int_\Omega \left(\sum_{n=0}^N X(n) \right)^2 dP$$
$$= \sum_{n=0}^\infty R(n) - \frac{R(0)}{2}.$$

A GENERALIZED **FDT** OF THE FIRST KIND ([32, 33]). On the basis of the first and second KMO-Langevin equations (4.15) and (4.16), we have the following:

(A-1)
$$\frac{1}{\beta_1(1+e^{i\theta}) + 1 - e^{i\theta} + 2\pi\tilde{\gamma}_1(\theta)} = \frac{h(e^{i\theta})}{2\lim_{\tau \downarrow -\pi} h(e^{i\tau})} \quad (\theta \in (-\pi, \pi)),$$

(A-2)
$$\frac{1}{\beta_2(1+e^{i\theta}) + 1 - e^{i\theta} + 2\pi\tilde{\gamma}_2(\theta)} = \frac{[R](e^{i\theta})}{2\lim_{\tau \downarrow -\pi} [R](e^{i\tau})} \quad (\theta \in (-\pi, \pi)).$$

A GENERALIZED **FDT** OF THE SECOND KIND ([32, 33]). On the basis of the first and second KMO-Langevin equations (4.15) and (4.16), we have the following:

(B-1)
$$\frac{\alpha_1^2}{2} = R(0) C_{\beta_1, \gamma_1},$$

(B-2)
$$\Delta_{\alpha_2 \mathbf{I}}(d\theta) = \frac{R(0)}{2\pi} \{ 2(1 + \beta_2 + \gamma_2(0)) \operatorname{Re}(\beta_2(1+e^{i\theta}) + 1 - e^{i\theta} + 2\pi\tilde{\gamma}_2(\theta))$$
$$- |\beta_2(1+e^{i\theta}) + 1 - e^{i\theta} + 2\pi\tilde{\gamma}_2(\theta)|^2 \} d\theta.$$

Here C_{β_1, γ_1} is a constant defined by

(6.6)
$$C_{\beta_1, \gamma_1} = \pi \left\{ \int_{-\pi}^\pi |\beta_1(1+e^{i\theta}) + 1 - e^{i\theta} + 2\pi\tilde{\gamma}_1(\theta)|^{-2} d\theta \right\}^{-1}$$

and $\Delta_{\alpha_2 \mathbf{I}}$ stands for the spectral measure of $\alpha_2 \mathbf{I}$.

A GENERALIZED EINSTEIN RELATION (1) ([32, 33]). On the basis of the first (resp., the second and the modified second) KMO-Langevin equations (4.15) (resp., (4.16) and (4.27)), the following relations hold, respectively:

(C-1) $$D = \frac{R(0)}{2\beta_1} \frac{C_{\beta_1,\gamma_1}}{2\beta_1},$$

(C-2) $$D = \frac{R(0)}{2\beta_2}(1 + \gamma_2(0)),$$

(C-2-m) $$D = \frac{R(0)}{\beta_{m,2}}.$$

Corresponding to inequality (6.4), we obtain

(6.7) $$\frac{C_{\beta_1,\gamma_1}}{2\beta_1} \geq 1.$$

Similarly, as in the continuous case, it follows that the equality in (6.7) holds only for the case $\rho_1 = 0$, that is, the weakly stationary process with Markovian property treated in Example 4.1. Therefore, we find in the discrete system that the classical Einstein relation (C) for the continuous system deviates on the basis of both the first and the second KMO-Langevin equations (4.15) and (4.16), but it is conserved on the basis of the modified second KMO-Langevin equation (4.27).

Finally, we have the following corresponding to relations (D) and (D-1) for the continuous system.

A GENERALIZED EINSTEIN RELATION (2) ([**32**]). On the basis of the first and the modified first KMO-Langevin equations (4.15) and (4.26), we have

(D-1) $$D = \frac{\alpha_1^2}{2(2\beta_1)^2},$$

(D-1-m) $$D = \frac{\alpha_{m,1}^2}{2(\beta_{m,1})^2}.$$

But neither the second nor the modified second KMO-Langevin equations (4.16) and (4.27) have these relations.

7. KM$_2$O-Langevin equation and FDT

Let $\mathbf{X} = (X(n); |n| \leq N)$ be an $\mathbb{R}^d$-valued weakly stationary process defined on a probability space $(\Omega, \mathcal{B}, P)$ with expectation vector 0. Here d, N are fixed natural numbers, and we do not assume that the process $\mathbf{X}$ has T-positivity. Denoting by R the correlation matrix function of the process $\mathbf{X}$, we define for each n, $1 \leq n \leq N$, a Toeplitz matrix S_n ($\in M(nd; \mathbb{R})$) by

(7.1) $$S_n = \begin{pmatrix} R(0) & R(1) & \cdots & R(n-1) \\ {}^t R(1) & R(0) & \cdots & R(n-2) \\ \vdots & \vdots & \ddots & \vdots \\ {}^t R(n-1) & {}^t R(n-2) & \cdots & R(0) \end{pmatrix}.$$

Note that

(7.2) $${}^t R(n) = R(-n).$$

We remark that under the condition

(7.3) $$R(0) \in \mathrm{GL}(d;\mathbb{R}),$$

either (7.4) or (7.5) holds:

(7.4) $$S_n \in \mathrm{GL}(nd;\mathbb{R}) \quad (1 \leq n \leq N),$$

(7.5) $$\begin{cases} S_n \in \mathrm{GL}(nd;\mathbb{R}) & (1 \leq n \leq N_0), \\ S_n \notin \mathrm{GL}(nd;\mathbb{R}) & (N_0 + 1 \leq n \leq N), \end{cases}$$

where $N_0 = \max\{n \in \{1,\ldots,N\}; \det(S_n) \neq 0\}$.

Recalling the notation (4.11), we define two $\mathbb{R}^d$-valued stochastic processes $v_+ = (v_+(n); 0 \leq n \leq N)$ and $v_- = (v_-(-n); 0 \leq n \leq N)$ by

(7.6) $\quad v_+(n) = X(n) - P_{\mathbf{M}_0^{n-1}(\mathbf{X})} X(n) \quad \text{and} \quad v_-(-n) = X(-n) - P_{\mathbf{M}_{-n+1}^0(\mathbf{X})} X(-n),$

where $P_{\mathbf{M}_n^m(\mathbf{X})}$ stands for the projection operator on the closed subspace $\mathbf{M}_n^m(\mathbf{X})$ ($n \leq m$) with the convention $\mathbf{M}_0^{-1}(\mathbf{X}) = \mathbf{M}_1^0(\mathbf{X}) = \{0\}$. Let $V_+(n)$ and $V_-(n)$ be the variance matrix of $v_+(n)$ and $v_-(-n)$, respectively. It then follows that

CAUSAL RELATION ([35, 40]). *For each n, m, $0 \leq n, m \leq N$,*

(7.7) $$v_+(0) = v_-(0) = X(0),$$

(7.8)
$$\int_\Omega v_+(n) {}^t v_+(m) \, dP = \delta_{nm} V_+(n) \quad \text{and} \quad \int_\Omega v_-(-n) {}^t v_-(-m) \, dP = \delta_{nm} V_-(n),$$

(7.9) $$\mathbf{M}_0^n(\mathbf{X}) = \mathbf{M}_0^n(v_+) \quad \text{and} \quad \mathbf{M}_{-n}^0(\mathbf{X}) = \mathbf{M}_{-n}^0(v_-).$$

In the sequel we shall treat the situation where condition (7.4) holds.

DECOMPOSITION THEOREM ([35, 39]). *There exists a unique system $\{\gamma_+(n,k), \gamma_-(n,k), \delta_+(m), \delta_-(m); 1 \leq k < n \leq N, 1 \leq m \leq N\}$ of $M(d;\mathbb{R})$ such that for each n, $1 \leq n \leq N$, the process $\mathbf{X}$ satisfies*

(7.10) $$X(n) = -\sum_{k=1}^{n-1} \gamma_+(n,k) X(k) - \delta_+(n) X(0) + v_+(n),$$

(7.11) $$X(-n) = -\sum_{k=1}^{n-1} \gamma_-(n,k) X(-k) - \delta_-(n) X(0) + v_-(-n).$$

DEFINITION 7.1. We shall call equation (7.10) (resp., 7.11)) the forward (resp., backward) KM$_2$O-Langevin equation associated with the process **X**. Further, the process v_+ (resp., v_-) and the system $\{\gamma_+(n,k), \gamma_-(n,k), \delta_+(m), \delta_-(m); 1 \leq k < n \leq N, 1 \leq m \leq N\}$ shall be called the forward (resp., the backward) KM$_2$O-Langevin force associated with the process **X** and the KM$_2$O-Langevin data associated with the correlation matrix function R.

REMARK 7.1. The KM$_2$O-Langevin equations were derived as the candidate for the discrete version to the $(\alpha,\beta,\gamma,\delta)$-Langevin equations which Miyoshi [18, 19] had derived in the continuous system.

Now we shall observe what kinds of representations **FDT**, stated in §6, possess on the basis of these KM$_2$O-Langevin equations.

FDT (1) ([**17, 4, 43, 45, 35, 40**]). For each n, k, $1 \leq k < n \leq N$, we have the following algorithms:

(7.12) $$\begin{cases} \gamma_+(n,k) = \gamma_+(n-1,k-1) + \delta_+(n)\gamma_-(n-1,n-k-1), \\ \gamma_-(n,k) = \gamma_-(n-1,k-1) + \delta_-(n)\gamma_+(n-1,n-k-1), \end{cases}$$

(7.13) $$\begin{cases} V_+(n) = (I - \delta_+(n)\delta_-(n))V_+(n-1), \\ V_-(n) = (I - \delta_-(n)\delta_+(n))V_-(n-1), \end{cases}$$

(7.14)
$$\delta_-(n)V_+(n-1) = V_-(n-1)\,{}^t\delta_+(n) \quad \text{and} \quad \delta_-(n)V_+(n) = V_-(n)\,{}^t\delta_+(n),$$

where $\gamma_+(n,0) = \delta_+(n)$ and $\gamma_-(n,0) = \delta_-(n)$.

Moreover, we shall investigate the covariance matrix function $I(n,m)$ between the forward KM$_2$O-Langevin force v_+ and the backward KM$_2$O-Langevin force v_-:

(7.15) $$I(n,m) = \int_\Omega v_+(n)\,{}^t v_-(-m)\, dP \qquad (0 \leq n, m \leq N).$$

FDT (2) ([**37**]). For each n, m, $1 \leq n \leq N$, $2 \leq m \leq N$, we have

(7.16) $$\begin{cases} I(n,0) = I(0,n) = 0, \\ I(n,1) = I(1,n) = -\delta_+(n+1)V_-(n), \\ I(n,m) = I(n+1, m-1) + \left(\sum_{k=1}^{m-2} I(n+1,k)\,{}^t\delta_+(k+1)\right){}^t\delta_-(m) \\ \qquad\qquad - \delta_+(n+1)\sum_{k=1}^{n-1} \delta_-(k+1)I(k,m). \end{cases}$$

It follows from (7.12)–(7.14) in **FDT** that the system $\{\delta_+(n); 0 \leq n \leq N\}$ determines other KM$_2$O-Langevin data and it can itself be calculated in terms of the correlation function R by (7.12)–(7.13) and the following algorithm:

ALGORITHM ([**17, 4, 43, 45, 35, 40**]). For each n, $1 \leq n \leq N$, we have

(7.17) $$\delta_+(n) = -\left(R(n) + \sum_{k=0}^{n-2} \gamma_+(n-1,k)R(k+1)\right)V_-(n-1)^{-1}.$$

By inverting the consideration above, assume that we are given any symmetric, positive definite matrix $V \in GL(d; \mathbb{R})$ and any N elements $\delta_+(n)$ of $M(d; \mathbb{R})$ ($1 \leq n \leq N$).

THE FIRST STEP. We construct a system $\{\gamma_+(n,k), \gamma_-(n,k), \delta_+(m), \delta_-(m), V_+(m), V_-(m); 1 \leq k < n \leq N, 1 \leq m \leq N\}$ in such a way that the relations (7.12)–(7.14) are inductively satisfied. We shall suppose in this step that all $V_+(m), V_-(m)$ ($0 \leq m \leq N-1$) are positive definite.

THE SECOND STEP. By using Kolmogorov's extension theorem, we construct two $\mathbb{R}^d$-valued stochastic processes $v_+ = (v_+(n); 0 \leq n \leq N)$ and $v_- = (v_-(-n); 0 \leq n \leq N)$ satisfying $v_+(0) = v_-(0)$ and relation (7.16) [**37**].

THE THIRD STEP. Under the initial condition $X(0) = v_+(0)$, we solve equations (7.10) and (7.11) to obtain an $\mathbb{R}^d$-valued stochastic process $\mathbf{X} = (X(n); |n| \leq N)$.

Thus, we can arrive at the following result:

CONSTRUCTION THEOREM ([35, 40, 37]). *The process* **X** *constructed above is a weakly stationary process with the system* $\{\gamma_+(n,k), \gamma_-(n,k), \delta_+(m), \delta_-(m), V_+(m), V_-(m); 1 \leq k < n \leq N, 1 \leq m \leq N\}$ *as its* KM_2O-*Langevin data.*

8. KM_2O-Langevin equations and $AR(\infty)$-Langevin equations

In this section we shall show that $AR(\infty)$-Langevin equations can be derived from KM_2O-Langevin equations treated in §7, by a limit procedure as $N \to \infty$ [36].

Let $\mathbf{X} = (X(n); n \in \mathbb{Z})$ be an $\mathbb{R}^d$-valued weakly stationary process with expectation vector 0 such that its spectral matrix measure has a continuous spectral density Δ defined on $[-\pi, \pi]$. We define a one-parameter unitary group $(U(n); n \in \mathbb{Z})$ operating on the Hilbert space $\mathbf{M}^\infty_{-\infty}(\mathbf{X})$ by

$$(8.1) \qquad U(n)(X_j(m)) = X_j(m+n)$$

and introduce for each $N \in \mathbb{N}$ two $\mathbb{R}^d$-valued weakly stationary processes $\varepsilon^N_+ = (\varepsilon^N_+(n); n \in \mathbb{Z})$ and $\varepsilon^N_- = (\varepsilon^N_-(n); n \in \mathbb{Z})$ with expectation vector 0 by

$$(8.2) \qquad \varepsilon^N_+(n) = U(n-N)v_+(N) \quad \text{and} \quad \varepsilon^N_-(n) = U(n+N)v_-(-N).$$

By using these processes, we find that the KM_2O-Langevin equation (7.10) (resp., (7.11)) associated with the local and weak stationary process $(X(n); |n| \leq N)$ can be transformed into the following generalized $AR(\infty)$-Langevin equation (8.3) (resp., (8.4)): for all $N \in \mathbb{N}$, $n \in \mathbb{Z}$,

$$(8.3) \qquad X(n) = -\sum_{k=1}^{N} \gamma_+(N, N-k) X(n-k) + \varepsilon^N_+(n),$$

$$(8.4) \qquad X(n) = -\sum_{k=1}^{N} \gamma_-(N, N-k) X(n+k) + \varepsilon^N_-(n).$$

We can use the method in Grenander-Szegö [9] to show

LEMMA 8.1. *There exist two positive definite matrices* V_+ *and* V_- $(\in GL(d; \mathbb{R}))$ *such that*

$$(8.5) \qquad \lim_{N \to \infty} V_+(N) = V_+ \quad \text{and} \quad \lim_{N \to \infty} V_-(N) = V_-.$$

By combining (7.12)–(7.13) in **FDT** with (8.5), we have

THEOREM 8.1. *There exist two* $\mathbb{R}^d$-*valued standard white noises* $\xi_+ = (\xi_+(n); n \in \mathbb{Z})$ *and* $\xi_- = (\xi_-(n); n \in \mathbb{Z})$ *such that*

$$(8.6) \qquad \underset{N \to \infty}{\text{l.i.m.}} \varepsilon^N_+(n) = (V_+)^{1/2} \xi_+(n) \quad \text{and} \quad \underset{N \to \infty}{\text{l.i.m.}} \varepsilon^N_-(n) = (V_-)^{1/2} \xi_-(n) \qquad (n \in \mathbb{Z}),$$

$$(8.7) \qquad \int_\Omega \xi_+(n)\, {}^t\xi_+(m)\, dP = \delta_{nm} I \quad \text{and} \quad \int_\Omega \xi_-(n)\, {}^t\xi_-(m)\, dP = \delta_{nm} I \qquad (n, m \in \mathbb{Z}).$$

Applying (8.3) and (8.4) to Theorem 8.1, we obtain

LEMMA 8.2. *There exists a unique system $\{\gamma_+(n), \gamma_-(n); n \in \mathbb{N}\}$ of $M(d; \mathbb{R})$ such that for each fixed $k \in \mathbb{N}$,*

(8.8) $$\lim_{N \to \infty} \gamma_+(N, N-k) = \gamma_+(k) \quad \text{and} \quad \lim_{N \to \infty} \gamma_-(N, N-k) = \gamma_-(k).$$

Letting $N \to \infty$ in (8.3) and (8.4), we can make use of the above results to obtain the following forward (resp., backward) AR(∞)-Langevin equation (8.9) (resp., (8.10)).

THEOREM 8.2.

(8.9) $$X(n) = -\operatorname*{l.i.m.}_{N \to \infty} \sum_{k=1}^{N} \gamma_+(N, N-k) X(n-k) + V_+ \xi_+(n) \quad \text{a.s.}$$

(8.10)
$$X(n) = -\operatorname*{l.i.m.}_{N \to \infty} \sum_{k=1}^{N} \gamma_-(N, N-k) X(n+k) + V_- \xi_-(n) \quad \text{a.s.}$$

Moreover, we have the forward causal relation among $\mathbf{X}, \xi_+$, and ξ_-.

THEOREM 8.3. *For each $n \in \mathbb{Z}$,*

(8.11) $$\mathbf{M}_{-\infty}^{n}(\mathbf{X}) = \mathbf{M}_{-\infty}^{n}(\xi_+) \quad \text{and} \quad \mathbf{M}_{n}^{\infty}(\mathbf{X}) = \mathbf{M}_{n}^{\infty}(\xi_-).$$

As an application of the results of this section, we shall finally lay stress on the fact that *a matrix outer function h of the spectral density matrix function Δ in the multidimensional case can be constructively obtained by the following formula.*

THEOREM 8.4. *For a.e. $\theta \in (-\pi, \pi]$,*

(8.12) $$h(e^{i\theta}) = \frac{1}{\sqrt{2\pi}} V_+ \left(1 + \lim_{N \to \infty} \sum_{k=1}^{N} \gamma_+(N, N-k) e^{ik\theta} \right)^{-1}.$$

9. AR-equations

We shall observe what position the AR-equations ($\equiv$ models), which are very often used in applied fields, occupy in the framework of the generalized AR-equations described in §8.

Let $\mathbf{X} = (X(n); n \in \mathbb{Z})$ be an $\mathbb{R}^d$-valued weakly stationary process with expectation vector 0 and correlation matrix function R satisfying condition (7.3). We shall fix any positive integer N.

DEFINITION 9.1. We say that $\mathbf{X}$ is an AR(K)-process if and only if there exist a system $\{A(k), \sigma; 1 \leq k \leq K\}$ of $M(d; \mathbb{R})$ and an $\mathbb{R}^d$-valued standard white noise $\xi_+ = (\xi_+(n); n \in \mathbb{Z})$ such that

(9.1) $$X(n) = \sum_{k=1}^{K} A(k) X(n-k) + \sigma \xi_+(n) \quad (n \in \mathbb{Z}),$$

(9.2) $$\sigma \in \operatorname{GL}(d; \mathbb{R}),$$

(9.3) $$\int_{\Omega} \xi_+(n)\,{}^t\xi_+(m)\, dP = \delta_{nm} I,$$

(9.4) $$\mathbf{M}_{-\infty}^{n}(\mathbf{X}) = \mathbf{M}_{-\infty}^{n}(\xi) \quad (n \in \mathbb{Z}).$$

We are now in a position to show

THEOREM 9.1 ([38]). *The following are equivalent*:
  (i) $\mathbf{X}$ *is an* $\mathrm{AR}(K)$-*process*.
  (ii) *Condition* (7.4) *holds for all* $N \in \mathbb{N}$, *and* $\delta_+(n) = 0$ *for all* $n \geq K + 1$.
  (iii) *Condition* (7.4) *holds for all* $N \in \mathbb{N}$, *and* $\mathbf{M}^n_{-\infty}(\varepsilon_+^K) = \mathbf{M}^n_{-\infty}(\mathbf{X})$ *for all* $n \in \mathbb{Z}$.

10. KM$_2$O-Langevin equations and causal relation

The stochastic processes represent mathematical models for random phenomena varying with time. It is to be noted that the meaning of the existence of causal relation among such processes has already appeared in (3.12), (3.15), (4.10), (4.14), (7.9), (8.11), and (9.4). In the sequel we shall consider the situation where the stochastic processes have discrete time parameter spaces [38].

DEFINITION 10.1. (i) For given $\mathbb{R}^{d_1}$- (resp. $\mathbb{R}^{d_2}$-) valued stochastic process $\mathbf{X}_1 = (X_1(n); n \in \mathbb{Z})$ (resp., $\mathbf{X}_2 = (X_2(n); n \in \mathbb{Z})$), we shall say that there exists a causal relation such that $\mathbf{X}_1$ is a causal and $\mathbf{X}_2$ is a result if and only if for any $n \in \mathbb{Z}$, there exists a Borel function F_n from $(\mathbb{R}^{d_1})^{\mathbb{N}^*}$ to $\mathbb{R}^{d_2}$ such that

$$(10.1) \qquad X_2(n) = F_n(X_1(n), X_1(n-1), \ldots) \quad \text{a.s.}$$

(ii) In particular, we shall say that the causal relation between $\mathbf{X}_1$ and $\mathbf{X}_2$ is linear (resp., nonlinear of pth order) according to whether the functions F_n can be taken as polynomials of the first degree (resp., p (≥ 2)th degree). In this case we shall write $\mathbf{X}_1 \xrightarrow{\text{linear}} \mathbf{X}_2$, $\mathbf{X}_1 \xrightarrow{\text{nonlinear of } p\text{th order}} \mathbf{X}_2$, respectively.

Putting $d_1 = d$ and $d_2 = 1$, we shall assume that the $\mathbb{R}^{d+1}$-valued stochastic process $\mathbf{X} = ({}^t({}^tX_1(n), X_2(n)); n \in \mathbb{Z})$ formed by an arrangement of $\mathbf{X}_1$ and $\mathbf{X}_2$ is weakly stationary with expectation vector 0. We define three kinds of correlation matrix functions R_1, R_2, and R_3 by

$$(10.2) \quad \begin{cases} R_1(n) = \int_\Omega X_1(n) {}^tX_1(0)\, dP \in M(d,d;\mathbb{R}), \\ R_2(n) = \int_\Omega X_2(n) {}^tX_1(0)\, dP \in M(1,d;\mathbb{R}), \\ R_3(n) = \int_\Omega X_2(n) X_2(0)\, dP \in M(1,1;\mathbb{R}). \end{cases}$$

Then we have a fundamental principle.

THEOREM 10.1. *The following are equivalent*:
  (i) $\mathbf{X}_1 \xrightarrow{\text{linear}} \mathbf{X}_2$.
  (ii) $C_n(\mathbf{X}_2|\mathbf{X}_1)$ *converges increasingly to* $R_3(0)$ *as* $n \to \infty$.
Here the function $C_n(\mathbf{X}_2|\mathbf{X}_1)$ stands for the causality function from $\mathbf{X}_1$ to $\mathbf{X}_2$ defined by

$$(10.3) \qquad C_n(\mathbf{X}_2|\mathbf{X}_1) \equiv \left(\int_\Omega (P_{\mathbf{M}_0^n(\mathbf{X}_1)} X_2(n))^2\, dP \right)^{1/2}.$$

For the purpose of calculating this causality function, we shall make use of the KM$_2$O-Langevin random force v_+ and the KM$_2$O-Langevin data $\{\gamma_+(n,k), V_+(k); 0 \leq k < n < \infty\}$ associated with the weakly stationary process $\mathbf{X}_1$. It follows from a local causal relation (7.9) between $\mathbf{X}_1$ and v_+ that there exists a system $\{C(n,k); 0 \leq k \leq n < \infty\}$ of $M(1,d;\mathbb{R})$ such that

$$(10.4) \qquad P_{\mathbf{M}_0^n(\mathbf{X}_1)} X_2(n) = \sum_{k=0}^{n} C(n,k) v_+(k) \qquad (n \in \mathbb{N}^*).$$

Thus, we can arrive at the following formula.

THEOREM 10.2. *For any n, k, $0 \leq k \leq n$,*

$$(10.5) \qquad C_n(\mathbf{X}_2|\mathbf{X}_1) = \left\{ \sum_{k=0}^{n} C(n,k) V_+(k)\, {}^t C(n,k) \right\}^{1/2},$$

$$(10.6)$$
$$C(n,k) = \begin{cases} R_2(n) R_1(0)^{-1} & (k=0), \\ \left\{ R_2(n-k) + \sum_{l=0}^{k-1} R_2(n-l)\, {}^t\gamma_+(k,l) \right\} V_+(k)^{-1} & (k \geq 1). \end{cases}$$

REMARK 10.1. In the case where $d = 1$, we can treat the problem of nonlinear causal relation of pth order from $\mathbf{X}_1$ to $\mathbf{X}_2$ by considering that of linear causal relation from $\mathbf{X}_3 = ({}^t(X_1(n), X_1(n)^2, \ldots, X_1(n)^p); n \in \mathbb{Z})$ to $\mathbf{X}_2$.

11. Stationary analysis and causal analysis

When we are given two kinds of data, we shall first consider the time series formed by them and apply our theory to them to propose two tests: Test(S) and Test(CS); Test(S) is a stationary test which checks the local and weak stationarity of them, and Test(CS) is a causal test which discriminates the existence and direction between two given kinds of data. Furthermore, we shall discuss their effectiveness [38, 39].

11.1. Stationary analysis. Suppose we are given $\mathbb{R}^d$-valued data $\mathscr{Z} = (\mathscr{Z}(n); 0 \leq n \leq N)$.

THE FIRST STEP. We define a sample mean vector $\mu^{\mathscr{Z}}$ and a sample correlation matrix function $R^{\mathscr{Z}} = (R_{jk}^{\mathscr{Z}}(*))_{1 \leq j,k \leq d}$ as follows:

$$(11.1) \qquad \mu^{\mathscr{Z}} \equiv \frac{1}{N+1} \sum_{n=0}^{N} \mathscr{Z}(n),$$

$$(11.2) \quad \begin{cases} R_{jk}^{\mathscr{Z}}(n) \equiv \dfrac{1}{N+1} \sum_{m=0}^{N-n} (\mathscr{Z}_j(n+m) - \mu_j^{\mathscr{Z}})(\mathscr{Z}_k(m) - \mu_k^{\mathscr{Z}}), \\ R_{jk}^{\mathscr{Z}}(-n) \equiv R_{kj}^{\mathscr{Z}}(n). \end{cases}$$

THE SECOND STEP. The standardized data $\mathscr{X} = {}^t(\mathscr{X}_1(n), \ldots, \mathscr{X}_d(n))$ of $\mathscr{Z}$ is defined by

$$(11.3) \qquad \mathscr{X}_j(n) \equiv (R_{jj}^{\mathscr{Z}}(0))^{-1/2}(\mathscr{Z}_j(n) - \mu_j^{\mathscr{Z}}) \qquad (1 \leq j \leq d,\ 0 \leq n \leq N).$$

We denote by $R^{\mathscr{X}}(n)$ ($0 \leq n \leq N$) its sample correlation matrix function. It follows from an experience rule in data analysis [1] that the maximal number M such that the sample correlation function $R^{\mathscr{X}}(n)$ $(0 \leq n \leq M)$ becomes our reliable data is given by

(11.4) $$M \equiv [3\sqrt{N+1}/d] - 1.$$

THE THIRD STEP. The function $R^{\mathscr{X}}(n)$ $(-M \leq n \leq M)$ obtained above can be extended as a nonnegative definite function, e.g., by letting $R^{\mathscr{X}}(n) = 0$ for $|n| \geq M+1$. In the same way as in the construction theorem (the first step) in §7, therefore, we can replace the function R in the algorithm (7.17) by the sample correlation function $R^{\mathscr{X}}$ to construct a sample KM$_2$O-Langevin data $\{\gamma_+(n,k), \gamma_-(n,k), \delta_+(m), \delta_-(m), V_+(m), V_-(m); 1 \leq k < n \leq M, 1 \leq m \leq M\}$ associated with $R^{\mathscr{X}}$. In this procedure we assume that all the matrices $V_+(m), V_-(m)$ $(0 \leq m \leq M-1)$ are positive definite.

THE FOURTH STEP. We shall replace $X(n)$ by $\mathscr{X}(n)$ in the forward KM$_2$O-Langevin equation (7.10) to construct a sample KM$_2$O-Langevin random force $v_+ = (v_+(n); 0 \leq n \leq M)$ by

(11.5) $$\begin{cases} v_+(0) \equiv \mathscr{X}(0), \\ v_+(n) \equiv \mathscr{X}(n) + \sum_{k=0}^{n-1} \gamma_+(n,k)\mathscr{X}(k) & (1 \leq n \leq M). \end{cases}$$

By taking $M+1$ elements $W_+(n)$ of $\mathrm{GL}(d;\mathbb{R})$ such that

(11.6) $$V_+(n) = W_+(n)\,{}^tW_+(n) \qquad (0 \leq n \leq M),$$

we shall proceed to a standardization of $\xi_+ = (\xi_+(n); 0 \leq n \leq M)$ of v_+:

(11.7) $$\xi_+(n) \equiv W_+(n)^{-1}v_+(n) = {}^t(\xi_{+1}(n), \ldots, \xi_{+d}(n)).$$

Moreover, we shall form $d(M+1)$ components of d-dimensional data ξ_+ into a line and construct a one-dimensional datum $\xi = (\xi(n); 0 \leq n \leq d(M+1)-1)$ as follows:

(11.8) $\quad \xi(n) \equiv \xi_{+j}(m), \qquad n = dm + j - 1 \qquad (1 \leq j \leq d,\ 0 \leq m \leq M).$

THE FUNDAMENTAL PRINCIPLE. By taking account of the decomposition theorem, **FDT**, and the construction theorem in §7, we find that the following (11.9) and (11.10) are equivalent each other:

(11.9) $\begin{cases} \mathscr{X} \text{ is a realization of a local and weak stationary process with } R^{\mathscr{X}} \\ \text{as its correlation matrix function.} \end{cases}$

(11.10) $\qquad \xi$ is a realization of a standard white noise.

THE FIFTH STEP. Since one set of original data $\mathscr{X}$ alone is insufficient for checking condition (11.9), we shall adopt $N - M + 1$ data $\mathscr{X}_i$ $(0 \leq i \leq N - M)$, which can be regarded as the same kind as the data $\mathscr{X}$, from the data $\mathscr{X}$ with length $N+1$ as follows:

(11.11) $$\mathscr{X}_i \equiv (\mathscr{X}(i+n); 0 \leq n \leq M).$$

Similarly, as ξ in (11.8), we shall construct for any i, $0 \leq i \leq N - M$, a one-dimensional datum $\xi_i = (\xi_i(n); 0 \leq n \leq d(M+1) - 1)$ by replacing $X(n)$ with $\mathscr{X}_i(n)$. It is our aim to investigate whether these data satisfy the fundamental principle (11.10). For that purpose, by making use of the central limit theorem and the law of large numbers, we shall propose three kinds of tests: $(M)_i, (V)_i$, and $(O)_i$. (M_i) is a test for mean 0, $(V)_i$ for variance 1, and $(O)_i$ for orthogonality; for each n, m, $1 \leq n \leq M$, $0 \leq m \leq M - n$,

$(M)_i$ $\qquad\qquad\qquad \sqrt{d(M+1)}|\mu^{\xi_i}| < 1.96,$

$(V)_i$ $\qquad\qquad\qquad |(v^{\xi_i} - 1)^\sim| < 2.2414,$

$(O)_i$ $\qquad\qquad\qquad d(M+1)\left(\sum_{j=1}^{2}(L_{n,m}^{(j)})^{-1/2}|R^{\xi_i}(n,m)|\right) < 1.96,$

where

(11.12) $\qquad\qquad \mu^{\xi_i} \equiv \dfrac{1}{d(M+1)} \displaystyle\sum_{k=0}^{d(M+1)-1} \xi_i(k),$

(11.13) $\quad (v^{\xi_i} - 1)^\sim \equiv \left\{\displaystyle\sum_{k=0}^{d(M+1)-1}(\xi_i(k)^2 - 1)\right\}\left\{\displaystyle\sum_{k=0}^{d(M+1)-1}(\xi_i(k)^2 - 1)^2\right\}^{-1/2},$

(11.14) $\qquad\qquad R^{\xi_i}(n,m) \equiv \dfrac{1}{d(M+1)} \displaystyle\sum_{k=m}^{d(M+1)-1-n} \xi_i(k)\xi_i(n+k).$

Moreover, the numbers $L_{n,m}^{(j)}$ are determined as follows: we denote by r the residue by dividing $d(M+1)$ by $2n$, and we set $[m/n] = s$.

(i) If $0 \leq r \leq n$,

(11.15) $\qquad \begin{cases} L_{n,m}^{(1)} = \begin{cases} n(q + s/2) - m & (s \text{ even}), \\ n(q - (s+1)/2) & (s \text{ odd}); \end{cases} \\ L_{n,m}^{(2)} = \begin{cases} n(q - 1 - s/2) + r & (s \text{ even}), \\ n(q - 1 + (s+1)/2) + r - m & (s \text{ odd}). \end{cases} \end{cases}$

(ii) If $n + 1 \leq r \leq 2n - 1$,

(11.16) $\qquad \begin{cases} L_{n,m}^{(1)} = \begin{cases} n(q - 1 + s/2) + r - m & (s \text{ even}), \\ n(q - 1 - (s+1)/2) + r & (s \text{ odd}); \end{cases} \\ L_{n,m}^{(2)} = \begin{cases} n(q - s/2) & (s \text{ even}), \\ n(q + (s+1)/2) - m & (s \text{ odd}). \end{cases} \end{cases}$

Note that these numbers $L_{n,m}^{(j)}$ satisfy the following:

(11.17) $\qquad\qquad\qquad d(M+1) - n - m = L_{n,m}^{(1)} + L_{n,m}^{(2)}.$

THE SIXTH STEP. We assert that the given data $\mathscr{X}$ is a realization of a local and weak stationary process when the following Test(S) is passed [39]:

(S) $\quad \begin{cases} \text{The rate of numbers } i \ (0 \leq i \leq N - M) \text{ such that} \\ \text{Test } ((M)_i), \text{Test } ((V)_i), \text{ and Test } ((O)_i) \\ \text{pass is more than 80\%, 70\%, and 80\%, respectively.} \end{cases}$

This is an experience rule which can be derived from a lot of experiments by using various kinds of data whose law and structure are well known. We regret not to be able to illustrate it mathematically. It can be obtained as a result of pushing applied mathematical study forward, with our definite object being to *check whether the data* has a local and weak stationarity, according to our guidance principle **FDP**. It seems that here lies importance, difficulty, and interest of applied mathematics.

11.2. Causal analysis. Finally, we shall apply the results in §10 to causal analysis. For data $\mathscr{Z}_1 = (\mathscr{Z}_1(n); 0 \leq n \leq N)$ formed by $N+1$ numbers of $\mathbb{R}^d$ and data $\mathscr{Z}_2 = (\mathscr{Z}_2(n); 0 \leq n \leq N)$ formed by $N+1$ numbers of $\mathbb{R}$, we shall consider the question of linear causality: $\mathscr{Z}_1 \xrightarrow{\text{linear}} \mathscr{Z}_2$.

THE SEVENTH STEP. We shall consider the situation where the $\mathbb{R}^{d+1}$-valued data $\mathscr{Z} = ({}^t({}^t\mathscr{Z}_1(n), \mathscr{Z}_2(n)); 0 \leq n \leq N)$ passes Test(S). We denote by

$$\mathscr{X} = ({}^t({}^t\mathscr{X}_1(n), \mathscr{X}_2(n)); 0 \leq n \leq N)$$

the standardized data of $\mathscr{Z}$. Note that the number M in equation (11.4) is $[3\sqrt{N+1}/(d+1)] - 1$ in this case. Furthermore, we denote by $R^{\mathscr{X}_1}, R^{\mathscr{X}_2;\mathscr{X}_1}$, and $R^{\mathscr{X}_2}$ three functions R_1, R_2, and R_3 in (10.2), respectively. Let the system $\{\gamma_+(n,k), \gamma_-(n,k), \delta_+(m), \delta_-(m), V_+(m), V_-(m); 1 \leq k < n \leq M, 1 \leq m \leq M\}$ be the same KM$_2$O-Langevin data associated with the data $R^{\mathscr{X}_1}$.

THE EIGHTH STEP. By replacing R_1 and R_2 in (10.5) and (10.6) with $R^{\mathscr{X}_1}$ and $R^{\mathscr{X}_2;\mathscr{X}_1}$, respectively, we shall draw up a graph of the function $C_n(\mathscr{X}_2|\mathscr{X}_1)$. Noting $R^{\mathscr{X}_2}(0) = 1$, we shall assert [38] that

(CS) $\begin{cases} \text{there exists a linear causality } \mathscr{Z}_1 \xrightarrow{\text{linear}} \mathscr{Z}_2 \text{ between both the data } \mathscr{Z}_1 \text{ and} \\ \text{the data } \mathscr{Z}_2 \text{ when the sample causality function } C_n(\mathscr{X}_2|\mathscr{X}_1) \ (0 \leq n \leq M) \\ \text{from } \mathscr{X}_1 \text{ to } \mathscr{X}_2 \text{ has a tendency to increase to 1.} \end{cases}$

REMARK 11.1. By using the same idea as in Remark 10.1, we can discuss the question of the existence of a nonlinear causal relation of pth order between the one-dimensional data $\mathscr{Z}_1$ and $\mathscr{Z}_2$.

12. Causal relation among meteorological data

12.1. We shall construct a dynamics by using some data whose rules are well known, and we shall investigate certain causal relations among them to observe the properties of sample causality functions.

For a one-dimensional datum $\mathscr{Z} = (\mathscr{Z}(n); 0 \leq n \leq 104)$, we shift it into

(12.1) $$\mathscr{Y}(n) = \mathscr{Z}(n+5) \quad (-5 \leq n \leq 99)$$

and define a dynamics made from five kinds of data $\mathscr{Z}_j = (\mathscr{Z}_j(n); 0 \leq n \leq 99)$ ($1 \leq j \leq 5$):

(12.2) $\begin{cases} \mathscr{Z}_1(n) = \mathscr{Y}(n), \quad \mathscr{Z}_2(n) = \mathscr{Y}(n)^2, \quad \mathscr{Z}_3(n) = \mathscr{Y}(n)^3, \\ \mathscr{Z}_4(n) = \mathscr{Y}(n-3) - 2\mathscr{Y}(n-5), \quad \mathscr{Z}_5(n) = \mathscr{Y}(n-3) - 2\mathscr{Y}(n-5)^2. \end{cases}$

As the data $\mathscr{Z}$, we adopt an AR(2)-data which can be obtained from the construction theorem in §7 by choosing

(12.3) $\quad (V, \delta(1), \delta(2)) = (1, 0.6, -0.3) \quad \text{and} \quad \delta(n) = 0 \quad (n \geq 3).$

By Theorem 9.1, this can also be described as follows. First we take a standardized uniform number $\xi = (\xi(n); 0 \le n \le 104)$ and then calculate the following:

$$\mathscr{Z}(n) = \begin{cases} -\sum_{k=0}^{n-1} \gamma(n,k)\mathscr{Z}(k) + \sqrt{V(n)}\xi(n) & (0 \le n \le 2), \\ -\sum_{k=1}^{2} \gamma(2, 2-k)\mathscr{Z}(n-k) + \sqrt{V(2)}\xi(n) & (3 \le n \le 104). \end{cases}$$

As the data ξ, we make use of 100 standardized uniform numbers with prime seed number from 2 to 541 and construct 10 kinds of two-dimensional data

$$({}^t(\mathscr{Z}_j(n), \mathscr{Z}_k(n)); 0 \le n \le 99) \qquad (1 \le j < k \le 5).$$

We show in Table 12.1 the results of Test(S) for them:

TABLE 12.1. Test(S)

	$\mathscr{Z}_2$	$\mathscr{Z}_3$	$\mathscr{Z}_4$	$\mathscr{Z}_4$
$\mathscr{Z}_2$	0.93	0.74	0.70	0.97
$\mathscr{Z}_3$		0.23*	0.99	0.58*
$\mathscr{Z}_4$			0.93	0.54*
$\mathscr{Z}_5$				0.97

TABLE 12.2. Test(S)

	Arct $\mathscr{Z}_3$	Arct $\mathscr{Z}_5$
Arct $\mathscr{Z}_3$	0.83	0.97
Arct $\mathscr{Z}_3$		0.98

We cannot find any desirable result in Test(S) for two-dimensional data formed by taking any pairings from $\mathscr{Z}_2, \mathscr{Z}_3$, and $\mathscr{Z}_5$. The reason for this seems to lie in the multiplication effect that they are made from taking the square and the cube of the data $\mathscr{Y}$. In order to remove it, we apply the arctangent transformation to each of them and define new data Arct $\mathscr{Z}_2$, Arct $\mathscr{Z}_3$, and Arct $\mathscr{Z}_5$. The results of Test(S) for these are shown in Table 12.2 with good results.

Since our data were shown to have local and weak stationarity, we can proceed to Test(CS). The results are illustrated in Figures 12.1–12.3. These teach us that the test of linear causality is useless for data that are combined with certain nonlinear relations and any test of nonlinear causality is also ineffective if the degree p of nonlinearity is different.

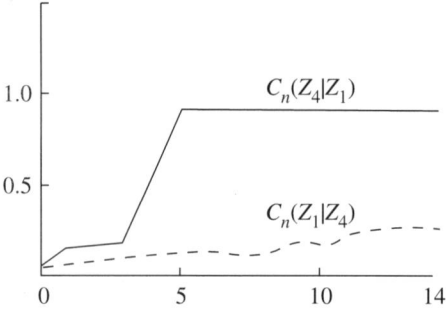

FIGURE 12.1. Test(CS)

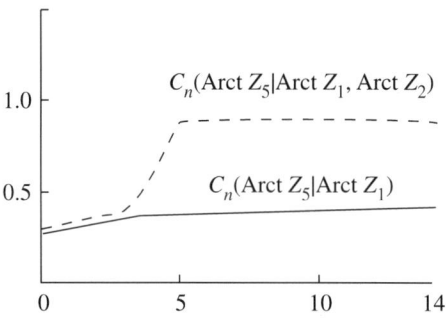

FIGURE 12.2. Test(CS)

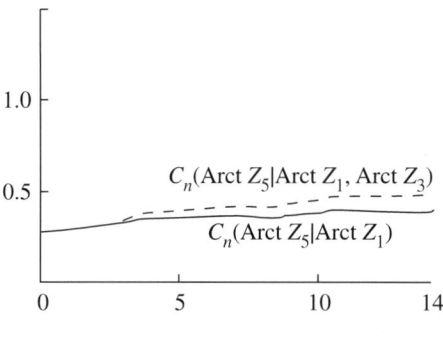

FIGURE 12.3. Test(CS)

12.2. Let $\mathscr{Z}_6 = (\mathscr{Z}_6(n); 0 \leq n \leq 113)$ and $\mathscr{Z}_7 = (\mathscr{Z}_7(n); 0 \leq n \leq 113)$ be Wolfer's sunspot numbers and the trapped numbers of Canadian lynx in the MacKenzie River from 1821 to 1934, respectively [**12, 46, 7**]. We show in Table 12.3 and Figure 12.4 the results of Test(S) and Test(CS) for two-dimensional data made from $\mathscr{Z}_6$ and $\mathscr{Z}_7$, respectively.

TABLE 12.3. Test(CS)

	(M)	(V)	(O)	(S)
${}^t(\mathscr{Z}_6, \mathscr{Z}_7)$	0.980	0.949	0.919	S

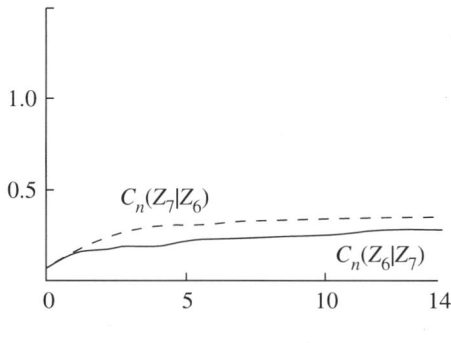

FIGURE 12.4. Test(CS)

We can obtain similar results also when we apply both logarithmic and arctangent transformations to the original data $\mathscr{Z}_6$ and $\mathscr{Z}_7$. Though these transformations alone are insufficient for certain objective assertions, we cannot say for certain from the comparison with Figures 12.1–12.3 that there exists a causal relation between Wolfer's sunspot numbers and the trapped numbers of Canadian lynx in the MacKenzie River.

12.3. Let $\mathscr{Z}_8 = (\mathscr{Z}_8(n); 0 \leq n \leq 99)$, $\mathscr{Z}_9 = (\mathscr{Z}_9(n); 0 \leq n \leq 99)$, and $\mathscr{Z}_{10} = (\mathscr{Z}_{10}(n); 0 \leq n \leq 99)$ be Wolfer's sunspot numbers, the rainfall average in a year in Sapporo, and the temperature average in a year in Sapporo from 1889 to 1988. We show the results of Test(S) for three kinds of two-dimensional data formed by these data.

The test of local and weak stationarity is rejected for the data $\mathscr{Z}_8$ and $\mathscr{Z}_{10}$. If we apply the arctangent transformation to them, however, we find that new data pass Test(S) whose results are shown in Table 12.5.

Therefore, we can proceed to Test(CS) for the data transformed by the arctangent transform. The results are illustrated in Figures 12.5–12.7.

TABLE 12.4. Test(S)

	(M)	(V)	(O)	(S)
$^t(\mathscr{Z}_8, \mathscr{Z}_9)$	0.977	0.860	1.000	S
$^t(\mathscr{Z}_8, \mathscr{Z}_{10})$	0.988	0.674*	0.988	NS
$^t(\mathscr{Z}_9, \mathscr{Z}_{10})$	1.000	0.872	0.988	S
$\mathscr{Z}_5$				0.97

TABLE 12.5. Test(S)

	(M)	(V)	(O)	(S)
$^t(\text{Arct } \mathscr{Z}_8, \text{Arct } \mathscr{Z}_9)$	1.000	1.000	1.000	S
$^t(\text{Arct } \mathscr{Z}_8, \text{Arct } \mathscr{Z}_{10})$	1.000	0.953	0.965	S
$^t(\text{Arct } \mathscr{Z}_9, \text{Arct } \mathscr{Z}_{10})$	0.988	0.907	0.977	S

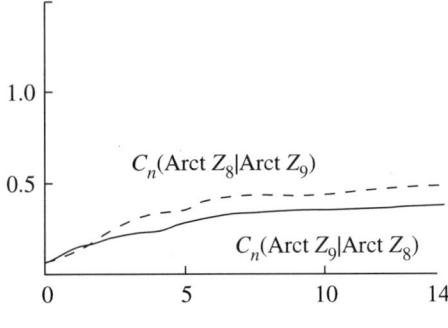

FIGURE 12.5. Test(CS)

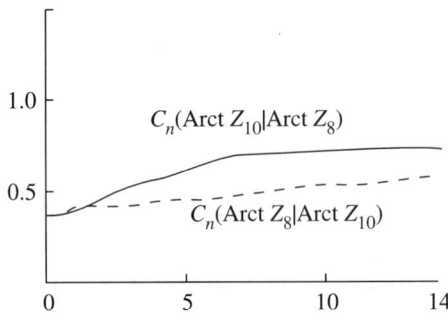

FIGURE 12.6. Test(CS)

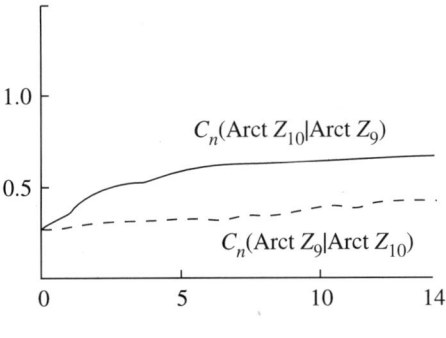

FIGURE 12.7. Test(CS)

It is likely that Wolfer's sunspot numbers cause the temperature in a given year in Sapporo and that the rainfall in a given year in Sapporo causes the temperature in a given year in Sapporo. However, we have to abstain from concluding from Figure 12.5 whether there exists a causal relation between Wolfer's sunspot numbers and the rainfall in a given year in Sapporo.

Finally, we would like to lay down our pen by stating some tasks that remain to be solved. It is very important for us to carry Test(CS) forward to give a systematic treatment for the nonlinear structure in random phenomena, because it relates to the settlement of nonlinear prediction problems by obtaining certain algorithms which are useful in applied fields. We are now engaged in such a study. Moreover, we intend to apply our theory to some important economic time series and compare the results on the basis of our theory with those based upon the causal analysis of economic time series due to Granger [8]. Furthermore, we think that the problem to quantify our decision standard for Test(CS) has to be overcome by all means in order that we can carry forward our project for analysis of time series under our philosophy **FDP**.

NOTES IN TRANSLATION. We have recently solved the problem of nonlinear prediction for one-dimensional strictly stationary processes in the following two papers: *Application of the theory of KM_2O-Langevin equations to the linear prediction problem for the multidimensional weakly stationary time series*, J. Math. Soc. Japan **45** (1993), 277–294 and *Application of the theory of KM_2O-Langevin equations to the nonlinear production problem for the one-dimensional strictly stationary time series*, to appear in J. Math. Soc. Japan, 1993.

References

1. H. Akaike and T. Nakagawa, *Statistical analysis and control of dynamics*, Science, 1972.
2. B. J. Alder and T. E. Wainwright, *Velocity autocorrelations for hard spheres*, Phys. Rev. Lett. **18** (1967), 988–990.
3. J. Boussinesq, *Sur la résistance qu'oppose un liquide indéfini en repos, sans pesanteur, au mouvement varié d'une sphère solide*, C. R. Acad. Sci. Paris **100** (1885), 935–937.
4. J. Durbin, *The fitting of time series models*, Rev. Internat. Statist. **28** (1960), 233–244.
5. H. Dym and H. P. McKean, Jr., *Gaussian processes, function theory and the inverse spectral problem*, Academic Press, New York, 1976.
6. A. Einstein, *Über die von der molekularkinetischen Theorie der Wärme geforderte Bewegung von in ruhenden Flüssigkeiten suspendierten Teilchen*, Drudes Ann. **17** (1905), 549–560.
7. C. Elton and M. Nicholson, *The ten-year cycle in numbers of the lynx in Canada*, J. Anim. Ecol. **11** (1942), 215–244.
8. C. W. J. Granger, *Investigating causal relations by econometric models and cross spectral methods*, Econometrica **37** (1969), 424–438.
9. U. Grenander and G. Szegö, *Toeplitz forms and their applications*, University of California Press, Berkeley and Los Angeles, 1958.
10. A. Inoue, *The Alder-Wainwright effect for stationary processes with reflection positivity*. I, J. Math. Soc. Japan **43** (1991), 515–526.
11. _____, *The Alder-Wainwright effect for stationary processes with reflection positivity*. II, Osaka J. Math. **28** (1991), 537–561.
12. W. S. Jevons, *Commercial crises and sun-spots*, Nature **15** (1878), 33–37.
13. R. Kubo, *Statistical mechanical theory of irreversible processes*. I: *General theory and simple applications to magnetic and conduction problem*, J. Phys. Soc. Japan **12** (1957), 570–586.
14. _____, *The fluctuation-dissipation theorem*, Rep. Progr. Phys. **29** (1966), 255–284.
15. _____, *Irreversible processes and stochastic processes*, The Theory of Stochastic Processes and Statistical Mechanics of Open System, R.I.M.S., vol. 367, Kôkyuroku, 1979, pp. 50–93.
16. P. Langevin, *Sur la théorie du mouvement brownien*, C. R. Acad. Sci. Paris **146** (1908), 530–533.
17. N. Levinson, *The Wiener RMS error criterion in filter design and prediction*, J. Math. Phys. **25** (1947), 261–278.
18. T. Miyoshi, *On (l,m)-string and $(\alpha,\beta,\gamma,\delta)$-Langevin equation associated with a stationary Gaussian process*, J. Fac. Sci. Univ. Tokyo, Sect. IA **30** (1983), 139–190.
19. _____, *On an $\mathbb{R}^d$-valued stationary Gaussian process associated with (k,l,m)-string and $(\alpha,\beta,\gamma,\delta)$-Langevin equation*, J. Fac. Sci. Univ. Tokyo Sect. IA **31** (1984), 154–194.
20. Y. Nakano and Y. Okabe, *On a multi-dimensional $[\alpha,\beta,\gamma]$-Langevin equation*, Proc. Japan Acad. **59** (1983), 171–173.
21. Y. Okabe, *On a stationary Gaussian process with T-positivity and its associated Langevin equation and S-matrix*, J. Fac. Sci. Univ. Tokyo Sect. IA **26** (1979), 115–165.

22. _____, *On a stochastic differential equation for a stationary Gaussian process with T-positivity and the fluctuation-dissipation theorem*, J. Fac. Sci. Univ. Tokyo Sect. IA **28** (1981), 169–213.
23. _____, *On a stochastic differential equation for a stationary Gaussian process with finite multiple Markovian property and the fluctuation-dissipation theorem*, J. Fac. Sci. Univ. Tokyo Sect. IA **28** (1982), 793–804.
24. _____, *On a Langevin equation*, Sûgaku **33** (1981), 306–324.
25. _____, *On a wave equation associated with prediction errors for a stationary Gaussian process*, Lecture Notes in Control and Inform. Sci., vol. 49, Springer, New York, 1983, pp. 215–226.
26. _____, *A generalized fluctuation-dissipation theorem for the one-dimensional diffusion process*, Commun. Math. Phys. **98** (1985), 449–468.
27. _____, *On KMO-Langevin equations for stationary Gaussian processes with T-positivity*, J. Fac. Sci. Univ. Tokyo Sect. IA. **33** (1986), 1–56.
28. _____, *On the theory of Brownian motion with the Alder-Wainwright effect*, J. Statist. Phys. **45** (1986), 953–981.
29. _____, *KMO-Langevin equation and fluctuation-dissipation theorem*. I, Hokkaido Math. J. **15** (1986), 163–216.
30. _____, *KMO-Langevin equation and fluctuation-dissipation theorem*. II, Hokkaido Math. J. **15** (1986), 317–355.
31. _____, *On long time tails of correlation functions for KMO-Langevin equations*, Proceedings of the Fourth Japan-USSR Symposium on Probability and Mathematical Statistics, Kyoto, Lecture Notes in Math., vol. 1299, Springer, Tokyo, 1986, pp. 391–397.
32. _____, *On the theory of discrete KMO-Langevin equations with reflection positivity*. I, Hokkaido Math. J. **16** (1987), 315–341.
33. _____, *On the theory of discrete KMO-Langevin equations with reflection positivity*. II, Hokkaido Math. J. **17** (1988), 1–44.
34. _____, *On the theory of discrete KMO-Langevin equations with reflection positivity*. III, Hokkaido Math. J. **18** (1989), 149–174.
35. _____, *On stochastic difference equations for the multi-dimensional weakly stationary time series*, Prospect of Algebraic Analysis (M. Kashiwara and T. Kawai, eds.), Academic Press, Tokyo, 1988, pp. 601–645.
36. _____, *The random forces associated with discrete-time weakly stationary processes*, Proceedings of Preseminar for International Conference on Gaussian Random Fields, Nagoya, 1991, pp. 126–134.
37. _____, KM_2O-*Langevin equation and the fluctuation-dissipation theorem* (to appear).
38. Y. Okabe and A. Inoue, *On the exponential decay of the correlation functions for KMO-Langevin equations*, Japan J. Math. **18** (1992), 13–24.
39. _____, *The theory of KM_2O-Langevin equations and its applications to data analysis*. II: *Causal analysis*, Nagoya Math. J. (to appear).
40. Y. Okabe and Y. Nakano, *The theory of KM_2O-Langevin equations and its applications to data analysis*. I: *Stationary analysis*, Hokkaido Math. J. **20** (1991), 45–90.
41. K. Oobayashi, T. Kohno, and H. Utiyama, *Photon correlation spectroscopy of the non-Markovian Brownian motion of spherical particles*, Phys. Rev. A **27** (1983), 2532–2641.
42. G. G. Stokes, *On the effect of the internal friction of fluids on the motion of pendulums*, Trans. Cambridge Philos. Soc. **9** (1856), 8–106.
43. P. Whittle, *On the fitting of multivariate autoregressions, and the approximate canonical factorization of a spectral density matrix*, Biometrika **50** (1963), 129–134.
44. A. Widom, *Velocity fluctuations of a hard-core Brownian motion*, Phys. Rev. A **3** (1971), 1394–1396.
45. R. A. Wiggins and E. A. Robinson, *Recursive solution to the multichannel fitting problem*, J. Geophys. Res. **70** (1965), 1885–1891.
46. G. U. Yule, *On a method of investigating periodicities in disturbed series, with Wolfer's sunspot numbers*, Philos. Trans. Roy. Soc. London Ser. A **226** (1927), 268–298.

Translated by YASUNORI OKABE

Analytic Capacity (A Theory of the Szegö Kernel Function)

Takafumi Murai

1. Introduction

1.1. Analytic functions. Analytic functions are huge interesting objects. One can understand only a small part by one method. Analytic function theory is a field theory from the viewpoint of flows and is a dynamical system from the viewpoint of vortex motion. The value-distribution theory of analytic functions is geometry, and the theory of special analytic functions is related to number theory. Recall $\sin z$, for example. This function has various meanings: (height)/(hypotenuse), (power series), (infinite product), $\operatorname{sn}(z, 0)$ (elliptic function), $y'' = -y$ (eigenfunction), $\frac{1}{i} \log \sin z$ (steady vortex potential), etc. Analyticity has two aspects; one is the fluid dynamical aspect based on the theory of incompressible and irrotational motion satisfying Euler's equation of motion, and the other is plane geometry deduced from $e^{i\theta} = \cos\theta + i\sin\theta$. The author believes that this double-faced aspect, together with naturality, yields numerous applications over the whole of mathematical science. Analytic capacity stated later has also various meanings and is a quantity concerning **the global structure** of analytic functions.

1.2. Analytic capacity and fluid dynamical background. The local structure of solutions of Cauchy-Riemann's equation (i.e., analytic functions, incompressible and irrotational 2-dimensional motions) reduces to the structure of the (global) fundamental solution (i.e., the Cauchy kernel, vortex filament) by Gauss-Green's formula (i.e., Cauchy's integral theorem). The study of ideal 2-dimensional velocity fields induced by a connected obstacle $E \subset \mathbf{C}$ (i.e., the study of incompressible and irrotational motions outside E) is nothing less than the study of a conformal mapping (i.e., Riemann's mapping) f_E from $E^c = \mathbf{C} \cup \{\infty\} - E$ onto the unit disk $\mathbf{D}$, and this mapping f_E is given by a solution of the following extremum problem: $\gamma(E) = \sup |f'(\infty)|$, where $f'(\infty) = \lim_{z \to \infty} z(f(\infty) - f(z))$ and the supremum is taken over all analytic functions f in E^c satisfying $|f| \leq 1$. The family of velocity fields induced by E is given by

$$\overline{g_{r,c,d}} = \overline{r(cf_E - \overline{c})(df_E + \overline{d})f_E^{-2}f_E'} \qquad (r \in \mathbf{R}, |c| = |d| = 1)$$

(cf. [**Mi1, Mi2**]), and the force (i.e., the lift) induced by $\overline{g_{r,c,d}}$ is equal to $C_0 r^2 (c^2 - d^2)/\gamma(E)$ by Blasius's formula, where C_0 is a constant depending only on

---

1991 *Mathematics Subject Classification.* Primary 30C85.
This article originally appeared in Japanese in Sûgaku **43** (4) (1991), 302–321.

the density of a given fluid and **R** denotes the real line. Thus, the study of the pair $(f_E, \gamma(E))$ is indispensable in fluid dynamics, and there are numerous articles about this pair. From the viewpoint of fluid dynamics, we now define analytic capacity. For a domain $\Omega \subset \mathbf{C} \cup \{\infty\}$ and $z \in \Omega$, we define analytic capacity by

$$c(z;\Omega) = \sup\{|f'(z)|; f \in H^\infty(\Omega), \|f\|_{H^\infty} \leq 1\},$$

where $H^\infty(\Omega)$ denotes the Banach space of bounded analytic functions in Ω with supremum norm $\|\cdot\|_{H^\infty}$. The analytic capacity of a compact set $E \subset \mathbf{C}$ is defined by $\gamma(E) = c(\infty; \Omega_E)$, where Ω_E is the connected component of E^c containing ∞. The function f_E satisfying $\gamma(E) = f'_E(\infty)$, $\|f_E\|_{H^\infty} = 1$ is the Ahlfors function. Our theme is to study the structure of the pair $(f_E, \gamma(E))$ so that the interference of several obstacles are clarified.

1.3. Szegö kernel function-theoretic background. The pair $(f_E, \gamma(E))$ has also various meanings and plays an important role in various areas. For the sake of simplicity, we explain the meaning of this pair assuming that

(1)   the boundary $\partial \Omega$ of $\Omega = \Omega_E$ consists of a finite number of analytic Jordan curves.

Let $H_0^2(\Omega)$ denote the Banach space of analytic functions f in Ω such that f is expressed as

$$f(z) = \frac{1}{2\pi} \int_{\partial \Omega} \frac{\mu(\zeta)}{\zeta - z} |d\zeta|$$

with a square integrable function μ on $\partial \Omega$ (with respect to $|d\zeta|$). The reproducing kernel $K(z, \overline{\zeta})$ of $H_0^2(\Omega)$ is the Szegö kernel function, i.e., $K(\infty, \overline{\zeta}) = \overline{K(\zeta, \overline{\infty})} = 0$,

$$f(z) = \frac{1}{2\pi} \int_{\partial \Omega} K(z, \overline{\zeta}) f(\zeta) |d\zeta| \qquad (f \in H_0^2(\Omega)).$$

The theory of the Szegö kernel function is a study of the global structure of analytic functions (i.e., a study of Riemann surfaces) and applicable to various areas: the theory of superstrings, the theory of Teichmüller spaces [**K**, Chapter 5] etc. This kernel function is ordinarily discussed together with a meromorphic function (i.e., the L-kernel function) $L(z, \zeta)$ determined by the following property [**Be**, Chapter 7]: For each $\zeta \in \Omega$, there exists a meromorphic function $L(z, \zeta)$ such that

(2) $\qquad L(z, \zeta) = \dfrac{1}{z - \zeta} + \text{(regular)}, \qquad L(z, \infty) = -L(\infty, z) = 0,$

(3) $\qquad \dfrac{1}{i} L(z, \zeta) dz = \overline{K(z, \overline{\zeta})} |dz| \quad \text{on } \partial \Omega,$

where the orientation of dz is chosen so that Ω lies to the left. The meromorphic function $L(z, \zeta)$ is an analytic continuation of $K(z, \overline{\zeta})$ to $\widetilde{\Omega} - \Omega$ ($\widetilde{\Omega}$ is the double-fold covering of Ω), and hence K is regarded as a meromorphic function in $\widetilde{\Omega}$ with a simple pole at ζ. The function

$$f(z, \zeta) = K(z, \overline{\zeta}) / L(z, \zeta)$$

is analytic in Ω (as a function of z), and $f(\Omega, \zeta)$ is an n-fold covering of $\mathbf{D}$ ($n = $ (the connectivity of Ω)) and

$$c(\zeta; \Omega) = f'(\zeta, \zeta) = K(\zeta, \overline{\zeta}).$$

Thus the study of the pair $(f(\cdot,\zeta), c(\zeta;\Omega))$ of a covering $f(\cdot,\zeta)$ and a metric $c(\zeta;\Omega)$ is nothing less than the study of the Szegö kernel function in $\widetilde{\Omega}$. The structure of this pair is equivalent to that of $(f_{\tau(E)}, \gamma(\tau(E)))$, where $\tau(z) = 1/(z-\zeta)$. Our theme in 1.2 is equivalent to clarifying the relation between $(f_E, \gamma(E))$ and the conformal geometric properties of Ω_E.

1.4. Construction-theoretic background. As is well known, "γ-null" is a concept useful in characterizing removable singularities in the sense of Riemann's theorem, and quantitative properties of γ play an important role in rational approximation [A1, V2], etc. The construction theory of $(f_E, \gamma(E))$ is indispensable in the search of quantitative properties, and our basic standpoint is to clarify the structure by constructing $(f_E, \gamma(E))$ concretely.

This article is not a general survey on analytic capacity but a summary of some topics with which the author is concerned. Analytic capacity is a big theme, and the following references are useful for the general theory, approximation theory, etc. [AB, G, Gar, Kor, O, Su4, SO, Z]. The references at the end of Grunski's lecture notes [Gr] are also useful. Finally, the author wishes to express his respects to the leaders in this subject in this country, Professors Yusaka Komatu, Nobuyuki Suita, and the late Professor Kotaro Oikawa on this occasion.

2. Basic formulae

2.1. Ahlfors-Garabedian's theory. In this chapter, we discuss some basic facts in the classical theory. The equality $\gamma(E) = 0$ is equivalent to the following property [A1]: *Each element of $H^\infty(\Omega - E)$ (Ω is a domain containing E) has an extension as an element of $H^\infty(\Omega)$*. Thus, $\gamma(\cdot) = 0$ characterizes the removable singularities for bounded analytic functions. Let $|\cdot|$ denote the 1-dimensional Hausdorff measure [F, p. 7], where $|\cdot|$ is normalized so that it is equal to the 1-dimensional Lebesgue measure as a set-function on **R**. The equality $|E| = 0$ implies $\gamma(E) = 0$ (Painlevé's Theorem). The Ahlfors function $f_E(\cdot) = f(\cdot;\Omega_E)$ exists uniquely. (The proof using the extremal point theorem in [Fi, p. 109] is simple.) For a while, we assume (1).

AHLFORS'S THEOREM ([A1]). $|f_E| = 1$ on $\partial\Omega$, $f_E(\infty) = 0$, $f_E(\Omega)$ is an n-fold covering of **D** ($n = $ (the connectivity of Ω)),

$$\log|f_E(z)| = -\sum_{j=1}^n G(z,\zeta_j), \qquad \sum_{j=1}^n \omega_k(\zeta_j) = 1 \quad (1 \leq k \leq n),$$

where $\{\zeta_j\}$ denotes the n zeros of f_E, $G(z,\zeta_j)$ is the Green's function with a pole at ζ_j and ω_k is the harmonic function (i.e., harmonic measure) such that $\omega_k = 1$ on Γ_k and $\omega_k = 0$ on $\partial\Omega - \Gamma_k (\partial\Omega = \bigcup_{j=1}^n \Gamma_j)$.

Let $H^p(\Omega)$ ($p \geq 1$) denote the Banach space of analytic functions ψ in Ω expressed as

$$\psi(z) = \psi(\infty) + \frac{1}{2\pi}\int_{\partial\Omega} \frac{\mu(\zeta)}{\zeta - z}|d\zeta|$$

for some pth power integrable function μ on $\partial\Omega$. The norm is defined by $\|\psi\|_{H^p} = \{\int_{\partial\Omega} |\psi|^p |dz|\}^{1/p}$.

GARABEDIAN'S DUALITY THEOREM ([**Ga1**]).

$$\gamma(E) = \inf\left\{\frac{1}{2\pi}\|\psi\|_{H^p}^p; \psi \in H^p(\Omega), \psi(\infty) = 1\right\}.$$

From this theorem, it is known that the study of γ is not only concerned with $H^\infty(\Omega)$ but also related to the theory of $H^p(\Omega)$ (cf. [**FS, Ga2, La, He2, Mu2**]). The solution $\psi_E = \psi(\cdot; \Omega_E)$ of the minimum problem for $p = 1$ is the Garabedian function. The Garabedian function ψ_E exists uniquely and $\log \psi_E$ is single-valued. The solution for $p > 1$ is $\psi_E^{1/p}$. The following relation exists between the Ahlfors function and the Garabedian function:

$$\frac{1}{i} f_E \psi_E \, dz = |\psi_E| \, |dz| \quad \text{on } \partial\Omega.$$

This relation corresponds to (3), and the equality $\psi_E(\infty) = 1$ corresponds to (2). The Garabedian function ψ_E for a general compact set E is defined by $\psi_E = \lim \psi(\cdot; \Omega_k)$, where $\{\Omega_k\}$ is an increasing sequence satisfying (1) such that $\infty \in \Omega_k \subset \Omega_E$ and $\bigcup \Omega_k = \Omega_E$ [**Sm, Su1, Su2**]. (The Garabedian function is defined independently of the choice of $\{\Omega_k\}$.)

The study of the Szegö kernel function is classical [**Be, Ga1, GS, Ha, HS, S, Sc1–Sc3, Su4, SS**]. (Note that the notation $\widehat{K}$ is used in [**Be**].) Now we show some properties related to γ. Equations (2) and (3) are equivalent to the following:

(4)
$$\begin{cases} K(z, \overline{\zeta}) = \dfrac{1}{2\pi} \displaystyle\int_{\partial\Omega} \dfrac{1}{w - z} \overline{L(w, \zeta)} |dw|, \\ L(z, \zeta) = \dfrac{1}{z - \zeta} + \dfrac{1}{2\pi} \displaystyle\int_{\partial\Omega} \dfrac{1}{w - z} \overline{K(w, \overline{\zeta})} |dw|, \\ K(\infty, \overline{\zeta}) = L(\infty, \zeta) = 0. \end{cases}$$

The following equation is another expression of the Szegö kernel function:

$$K(z, \overline{\zeta}) = \sum_{k=1}^\infty u_k(z) \overline{u_k(\zeta)},$$

where $\{u_k\}$ is a complete orthonormal basis of $H_0^2(\Omega)$. A system (4) of equations yields the following:

$$K(z, \overline{\zeta}) = \overline{K(\zeta, \overline{z})}, \qquad L(z, \zeta) = -L(\zeta, z).$$

In order to see the relation with $\gamma(E) = c(\infty; \Omega_E)$, we introduce the following two functions:

(5) $$g(z; \Omega) = f(z; \Omega)\phi(z; \Omega), \qquad \phi(z; \Omega) = \sqrt{\psi(z; \Omega)}.$$

Then

(6)
$$\begin{cases} g(z; \Omega) = \dfrac{1}{2\pi} \displaystyle\int_{\partial\Omega} L(\zeta, z) |d\zeta|, \\ \phi(z; \Omega) = 1 - \dfrac{1}{2\pi} \displaystyle\int_{\partial\Omega} \overline{K(\zeta, \overline{z})} |d\zeta|. \end{cases}$$

The reproducing kernel function of $H^2(\Omega)$ is given by

$$\frac{1}{\gamma(E)} \phi(z; \Omega) \overline{\phi(\zeta; \Omega)} + K(z, \overline{\zeta}).$$

By (6), it follows that

$$\gamma(E) = (2\pi)^{-2} \int_{\partial\Omega} \int_{\partial\Omega} K(z,\overline{\zeta}) \, dz \, \overline{d\zeta}. \tag{7}$$

The Szegö kernel function of a domain satisfying (1) and not containing infinity is defined analogously. The Szegö kernel function of a general domain is defined by using an increasing sequence of domains satisfying (1) [**Su1, Su2**]. The relation with automorphic functions and ϑ functions is seen in [**Fa**], and the relation with the Bergman kernel function $\partial^2 G(z,\zeta)/\partial z \overline{\partial \zeta}$ is given in [**He1, SH**].

2.2. Calculus. The computational approach to $(f_E, \gamma(E))$ is indispensable for understanding the structure of this pair. In this section, we show how to compute $\gamma(E)$. From the viewpoint of the computation, it is useful to rewrite Ahlfors-Garabedian's theorem in the following form.

THEOREM 1. *Let E be a compact set satisfying* (1), *and let* (f_0, ψ_0), $f_0 \in H^\infty(\Omega)$, $\psi_0 \in H^1(\infty)$ $(\Omega = \Omega_E)$ *be a pair satisfying the following two conditions*:

$$f_0(\infty) = 0, \quad |f_0| = 1 \quad \text{on } \partial\Omega, \quad \psi_0(\infty) = 1, \tag{8}$$

$$\frac{1}{i} f_0 \psi_0 \, dz = |\psi_0||dz| \quad \text{on } \partial\Omega. \tag{9}$$

Then

$$\gamma(E) = f'_0(\infty) = \frac{1}{2\pi} \|\psi_0\|_{H^1}. \tag{10}$$

In the proof, the following equality is important:

$$\gamma(E) = \sup\{|f'(\infty)|; \|f\|_{H^\infty} \leq 1, f(\infty) = 0\}. \tag{11}$$

In fact, for $f \in H^\infty(\Omega)$ such that $\|f\|_{H^\infty} \leq 1$, we form

$$g(z) = (f(z) - f(\infty))/(1 - \overline{f(\infty)} f(z)).$$

Then $g(\infty) = 0$, $|g'(\infty)| \geq |f'(\infty)|$, and $\|g\|_{H^\infty} \leq 1$, which yields (11). Now the proof of (10) is given in five lines. For the pair (f_0, ψ_0), the following inequalities hold:

$$|f'_0(\infty)| \leq \gamma(E) = \sup_{|f|\leq 1, f(\infty)=0} |f'(\infty)|$$

$$= \sup_{|f|\leq 1, f(\infty)=0} \left| \frac{1}{2\pi} \int_{\partial\Omega} f \psi_0 \, dz \right| \quad (\text{by } \psi_0(\infty) = 1)$$

$$\leq \frac{1}{2\pi} \|\psi_0\|_{H^1} = \frac{1}{2\pi} \int_{\partial\Omega} |f_0 \psi_0| \, |dz| \quad (\text{by } |f_0| = 1 \text{ on } \partial\Omega)$$

$$= \frac{1}{2\pi i} \int_{\partial\Omega} f_0 \psi_0 \, dz \quad (\text{by (9)})$$

$$= f'_0(\infty) \quad (\text{by } \psi_0(\infty) = 1).$$

Thus, all inequalities are equalities, which implies (10). From Ahlfors-Garabedian's Theorem, the pair (f_0, ψ_0) turns out to be (f_E, ψ_E); however, it is important to notice that $\gamma(E)$ is computed if the pair (f_0, ψ_0) has been found. There are various methods of constructing (f_E, ψ_E); the method using Green's function [**N**, p. 378], the method by conformal mappings [**Be**, p. 79], the method using Hahn-Banach's Theorem [**Gar,**

p. 19], etc. In the case where E consists of a finite number of analytic arcs, we take the boundary $\partial\Omega_E$ having two sides. Then the same argument works. Here are some examples:

(12) $$\gamma(\overline{\mathbf{D}}) = 1,$$

(13) $$\gamma(E) = |E|/4 \quad (E \subset \mathbf{R}) \quad [\mathbf{P}],$$

(14) $$\gamma(E) = \sin(|E|/4) \quad (E \subset \partial\mathbf{D}) \quad [\mathbf{B}, \mathbf{Mu3}],$$

$$\gamma(\overline{D(-c,r) \cup D(c,r)}) = \frac{2}{\pi} kK(k) \tanh\left(\frac{\pi}{2} \frac{K(k')}{K(k)}\right) \quad (0 < r < c),$$

where $D(c, r)$ denotes the open disk with center c and radius r, K is the complete elliptic integral of the first kind, and the modulus k is defined by

$$\frac{c}{r} = \cosh\left(\frac{\pi}{2} \frac{K(k')}{K(k)}\right) \quad (k' = \sqrt{1-k^2}).$$

We have (12) by setting $f_0 = -1/z$, $\psi_0 = 1$, and we obtain (13) by setting

$$f_0 = \left\{\exp\left(-\frac{\pi}{4}\mathscr{C}\chi_E\right) - \exp\left(\frac{\pi}{4}\mathscr{C}\chi_E\right)\right\} / \left\{\exp\left(-\frac{\pi}{4}\mathscr{C}\chi_E\right) + \exp\left(\frac{\pi}{4}\mathscr{C}\chi_E\right)\right\},$$

$$\psi_0 = \frac{1}{4}\left\{\exp\left(-\frac{\pi}{4}\mathscr{C}\chi_E\right) + \exp\left(\frac{\pi}{4}\mathscr{C}\chi_E\right)\right\}^2, \quad \mathscr{C}\chi_E(z) = \frac{1}{\pi}\int_E \frac{dt}{t-z}.$$

Analytic capacity γ is conformally invariant in the following sense: For a conformal mapping h from Ω_E onto Ω_F of form $h(z) = az + a_0 + a_1/z + \cdots (z \to \infty)$, the equality $\gamma(F) = |a|\gamma(E)$ holds. If E is connected, then $\gamma(E) = \operatorname{Cap}(E)$, where $\operatorname{Cap}(\cdot)$ is logarithmic capacity [$\mathbf{Z}$, p. 134]. Thus, the tables in [$\mathbf{L}$, pp. 165–167], [$\mathbf{R}$, pp. 348–351] are applicable to analytic capacity for continua. As an estimate of analytic capacity from the above, the following inequality is known [$\mathbf{P}$]:

$$\gamma\left(\bigcup_{k=1}^n E_k\right) \leq \gamma\left(\left\{\sum_{k=1}^n z_k ; z_k \in E_k \ (1 \leq k \leq n)\right\}\right).$$

In the case where $E \subset \mathbf{R} \cup (i\mathbf{R})$, the form of $\gamma(E)$ is very complicated. (The author tried many times to get a comprehensible expression without success.) The above examples show that the relation between $(f_E, \gamma(E))$ and the planar geometry of E is complicated. On the other hand, (13) shows that the structure of $(f_E, \gamma(E))$ is relatively simple in the case where Ω_E is a Denjoy domain (i.e., Ω_E is a domain whose complement is contained in $\mathbf{R}$.)

3. The Hilbert transform

3.1. Fundamental theory. In order to study analytic capacity, it is sufficient to confine ourselves to compact sets E satisfying (1). Given a compact set E satisfying (1), we can remove a finite number of arcs $\{\Lambda_k\}$ from $\partial\Omega$ ($\Omega = \Omega_E$) so that $\partial\Omega - \bigcup \Lambda_k$ consists of a finite number of analytic arcs and $\gamma(\partial\Omega) - \gamma(\partial\Omega - \bigcup \Lambda_k)$ is less than a given positive number. (The equality $\gamma(E) = \gamma(\partial\Omega)$ is important in this procedure.) Thus, we can restrict our attention to compact sets consisting of a finite number of analytic arcs. The totality of such compact sets is written by $\mathscr{A}$. For $E \in \mathscr{A}$, let

$L^p(E)$ $(1 \leq p \leq \infty)$ denote the L^p-space of functions on E with respect to $|dz|$. The norm is denoted by $\|\cdot\|_p$. The Hilbert transform $\mathscr{H} = \mathscr{H}_E$ is defined by

$$\mathscr{H}\mu(z) = \frac{1}{\pi} \text{ p.v.} \int_E \frac{1}{\zeta - z} \mu(\zeta)|d\zeta| \qquad (z \in E, \mu \in L^p(E)),$$

where p.v. denotes Cauchy's principal value. The Hilbert transform plays an important role in the study of $\gamma(E)$, $E \in \mathscr{A}$. Roughly speaking, the Hilbert transform enables us to rewrite various problems about analytic capacity as integral equations on $\mathscr{H}$. In this section, we rewrite Ahlfors-Garabedian's Theorem in terms of $\mathscr{H}$. The Hilbert transform $\mathscr{H}$ is bounded from $L^p(E)$ $(p > 1)$ to itself; the norm is denoted by $\|\mathscr{H}\|_{p,p}$. The Hilbert transform is $(L^1(E)$, weak $L^1(E))$-bounded, i.e., there exists a constant C such that

$$|\{z \in E; |\mathscr{H}\mu(z)| > \lambda\}| \leq C\|\mu\|_1/\lambda \qquad (\lambda > 0, \mu \in L^1(E)).$$

The minimum of such constants C is denoted by $\|\mathscr{H}\|_{1,w}$. In the case where $E = \mathbf{R}$, $\mathscr{H}$ is the standard Hilbert transform and $\mathscr{H}\mu + i\mu$ $(\mu \in L^1(\mathbf{R}))$ has an analytic extension to the upper half-plane. The Cauchy transform of $\mu \in L^1(E)$ is defined by

$$\mathscr{C}\mu(z) = \frac{1}{\pi} \int_E \frac{1}{\zeta - z} \mu(\zeta)|d\zeta| \qquad (z \in E^c).$$

The operator $\text{Id} - \mathscr{H}\overline{\mathscr{H}}$ from $L^2(E)$ to itself is invertible, where Id is the identity operator and $\overline{\mathscr{H}}$ is defined by $\overline{\mathscr{H}}\mu = \overline{\mathscr{H}\overline{\mu}}$. The inverse operator is denoted by $\mathscr{T}_E$.

THEOREM 2. $\gamma(E) = \frac{1}{\pi}\int_E \mathscr{T}_E 1 |dz|$.

In this theorem, the method of the proof itself is more essential than the assertion. The boundary $\partial\Omega$ of a domain $\Omega = \Omega_E$ is a double-fold covering of E; the points on $\partial\Omega$ corresponding to the endpoints of E are regarded as single. Let $E_{\text{en}} = \{\text{endpoints}\}$, $E^{\text{in}} = E - E_{\text{en}}$, $\partial_0\Omega = \partial\Omega - E_{\text{en}}$, and let $z_\pm$ denote two points on $\partial_0\Omega$ corresponding to $z \in E^{\text{in}}$. The boundary value of $\mathscr{C}\mu$ at $\zeta \in \partial_0\Omega$ is given by

$$\mathscr{H}\mu(\tilde{\zeta}) + i\mu(\tilde{\zeta})\frac{|d\zeta|}{d\zeta},$$

where $\tilde{\zeta}$ is the projection of $\zeta \in \partial_0\Omega$ to E^{in}. Now we pay attention to the fact that the pair $(g(\cdot;\Omega),\phi(\cdot;\Omega))$ defined by (5) is characterized by the following:

(15) $$\begin{cases} g(\infty;\Omega) = 0, \quad \phi(\infty;\Omega) = 1, \\ \frac{1}{i}\phi(z;\Omega)\,dz = \overline{g(z;\Omega)}|dz| \quad \text{on } \partial_0\Omega \end{cases}$$

(see (8), (9)). Put $g(\cdot;\Omega) = \mathscr{C}\overline{\mu}$, $\phi(\cdot;\Omega) = 1 + \mathscr{C}v$ $(\mu, v \in L^2(E))$. Then (15) is rewritten as

$$\frac{1}{i}\left\{1 + \mathscr{H}v + iv\frac{|dz|}{dz}\right\}dz = \left\{\overline{\mathscr{H}}\mu - i\mu\frac{dz}{|dz|}\right\}|dz| \quad \text{on } \partial_0\Omega.$$

In order to solve this equation, it is sufficient to solve the following system of equations which does not contain dz: $v = \overline{\mathscr{H}}\mu$, $1 + \mathscr{H}v = \mu$. Eliminating v, we obtain

(16) $$(\text{Id} - \mathscr{H}\overline{\mathscr{H}})\mu = 1.$$

From $g'(\infty;\Omega) = \frac{1}{\pi}\int_E \mu|dz|$, Theorem 2 is deduced.

This method is concrete, because the solution $\mu_E \in L^2(E)$ of (16) satisfies
$$g(\cdot;\Omega) = \mathscr{C}\overline{\mu}_E, \qquad \phi(\cdot;\Omega) = 1 + \mathscr{C}(\overline{\mathscr{H}\mu_E}).$$
The construction of μ_E is as follows: Fixing a positive number R_0 so that $R_0 > \|\mathscr{H}\|_{2,2}^2$, we define a sequence $(\mu_m)_{m=0}^\infty$ of functions inductively by $\mu_0 = 0$,
$$(R_0 + 1)\mu_m - \mathscr{H}\overline{\mathscr{H}}\mu_m = R_0\mu_{m-1} + 1 \qquad (m \geq 1).$$
Note that the explicit construction of the inverse operator of $(R_0 + 1)\,\mathrm{Id} - \mathscr{H}\overline{\mathscr{H}}$ is possible. The skew-symmetricity of $\mathscr{H}$ shows that $(v, \mathscr{H}\overline{\mathscr{H}}v) \geq 0$ ($v \in L^2(E)$), which yields the convergence of μ_m in $L^2(E)$. Thus $\mu_E = \lim_{m \to \infty} \mu_m$.

3.2. Balayage equation for $\mathscr{H}$. We now show a basic formula on $\mathscr{H}$ to handle Theorem 2. In terms of $\mathscr{H}$, analyticity is rewritten as

(17) $\quad \mathscr{H}\{\mu\mathscr{H}v + \mathscr{H}\mu \cdot v\} = \mathscr{H}\mu \cdot \mathscr{H}v - \tau\mu v \qquad (\mu, v \in L^2(E))$,

where $\tau = \overline{dz_+}/dz_+$. The balayage equation for $\mathscr{H}$ with respect to $c \in \mathbf{C}$, $h \in L^\infty(E)$ is defined by

(18) $\qquad\qquad \mathscr{H}\mu = c\overline{\rho}\mu + h$,

where $\rho = dz_+/|dz|$. (Note that τ is independent of the choice of $z_\pm$ and that ρ depends on the choice of $z_\pm$.) Our method for the study of (16) and (35) in §6 is to construct the solutions of the balayage equations concretely. Let $\{\Gamma_k\}_{k=1}^n$ denote the components of $E \in \mathscr{A}$, and let a_k, b_k denote the endpoints of each Γ_k. We redefine $z_\pm \in \partial_0\Omega$ so that
$$\int_{\Gamma_k} dz_+ = b_k - a_k \qquad (1 \leq k \leq n).$$
For $c \in \mathbf{C}$, $c \neq \pm i$, we set
$$\lambda_c(z) = \exp\left\{\left(\alpha + \frac{1}{2}\right)\,\mathrm{p.v.}\int_E \frac{d\zeta_+}{\zeta - z}\right\},$$
$$\lambda_c^*(z) = \lambda_c(z) \prod_{k=1}^n \frac{1}{z - b_k},$$
where $\alpha = -\frac{1}{\pi}\arctan c$, $-\frac{1}{2} \leq \Re\alpha < \frac{1}{2}$. We write $\mathbf{L} = \{iy;\, y \in \mathbf{R}, |y| \geq 1\}$. By combining (17) with the equality $\mathscr{H}(\lambda_c\rho) = c\lambda_c - \sec\pi\alpha$, we obtain the following theorem.

THEOREM 3. (19) *Let $c \in \mathbf{C} - \mathbf{L}$, $h \in L^\infty(E)$. Then $\mu \in L^1(E)$ is a solution of* (18) *if and only if*
$$\mu = -\frac{1}{1+c^2}\lambda_c\rho\mathscr{H}(h\rho/\lambda_c) - \frac{c}{1+c^2}h\rho + P\lambda_c^*\rho$$
for a polynomial P of degree less than $n = $ (the number of components of E).

(20) *Let $c \in \mathbf{L}$, $c \neq \pm i$, $h \in L^\infty(E)$. Then*
$$-\frac{1}{1+c^2}\lambda_c\rho\mathscr{H}(h\rho/\lambda_c) - \frac{c}{1+c^2}h\rho$$
is the unique solution of (18).

In the case where $c = \pm i$, there exists no solution of (18) in general. By (19), the solution of (18) is not unique in the case where $c = 1$. Thus, in the definition of the pair $(g(\cdot; \Omega), \phi(\cdot; \Omega))$, the condition "$\mu_E \in L^2(E)$" is indispensable. Finally, we give an example of the computation of $\gamma(E)$ by Theorem 2. Formula (17) shows that for $E \subset \mathbf{R}$,

$$(m+1)\mathscr{H}_E^{(m+1)}1 = \mathscr{H}_E 1 \cdot \mathscr{H}_E^{(m)}1 - m\mathscr{H}_E^{(m-1)}1 \qquad (m \geq 1),$$

where $\mathscr{H}_E^{(0)} = \mathrm{Id}$ and $\mathscr{H}_E^{(m)}$ is the composition of m copies of $\mathscr{H}_E$. This equality immediately yields that

$$\int_E \mathscr{H}_E^{(2m)} 1 |dz| = \frac{(-1)^m}{2m+1} |E| \qquad (m \geq 0).$$

Combined with Leibniz's formula, Theorem 2 shows that

$$\gamma(E) = \frac{|E|}{\pi} \sum_{m=0}^{\infty} \frac{(-1)^m}{2m+1} = \frac{|E|}{4}.$$

This is our interpretation of $\frac{1}{4}$ in (13) [**Mu2**]. Recall that Koebe's one quarter theorem is deduced from this $\frac{1}{4}$ [**A2**, p. 29].

3.3. Norm of $\mathscr{H}$. There are many articles about γ from the point of view of the functional analytic method [**G, Gar, Ha, Ma, DØ**]. In this section, we show the relation between γ and $\|\mathscr{H}\|_{1,w}$. For $E \in \mathscr{A}$, we set

$$\gamma_+(E) = \sup \left\{ \frac{1}{\pi} \|\mu\|_1; \ \mu \in L^1(E), \ \mu \geq 0, \ \|C\mu\|_{H^\infty} \leq 1 \right\}.$$

Evidently, $\gamma_+(E) \leq \gamma(E)$. We define two set functions by

$$\rho(E) = \inf \gamma(K)/|K|, \qquad \rho_+(E) = \inf \gamma_+(K)/|K|,$$

where the infimum is taken over all compact sets K on E. The following separation theorem plays a basic role in the functional analytic method:

SEPARATION THEOREM. *Let X, Y be two compact convex sets in the Banach space $C(E)$ of continuous functions on E. Then there exists a measure ξ on E such that*

$$\Re \int_E \mu \, d\xi > \Re \int_E \nu \, d\xi \qquad (\mu \in X, \nu \in Y)$$

(*see* [**Ch**, *p.* 102; **DØ**; **J1**, *p.* 49; **Mu1**, *p.* 74]).

From this separation theorem, we obtain the following

THEOREM 4 ([**Mu1**, p. 72]).

$$C_1/\|\mathscr{H}_E\|_{1,w} \leq \rho_+(E) \leq C_2/\|\mathscr{H}_E\|_{1,w}, \qquad \rho_+(E) \leq \rho(E) \leq C_2 \rho_+(E)^{1/3},$$

where C_1, C_2 are absolute constants.

COROLLARY 5. $\gamma(K) \geq C_1 |K|/\|\mathscr{H}_E\|_{1,w} \ (K \subset E)$.

This corollary shows that $\gamma(E)$ is comparable with $1/\|\mathscr{H}\|_{1,w}$. Let BMO denote the Banach space of functions on $\mathbf{R}$ (mod constants) of bounded mean oscillation. The norm is defined by

$$\|\mu\|_{BMO} = \sup_{I \text{ interval}} \frac{1}{|I|} \int_I \left| \mu(x) - \left(\frac{1}{|I|} \int_I \mu(y)\,dy \right) \right| dx.$$

Note that $L^\infty(\mathbf{R}) \subset \text{BMO}$. For a real-valued function $a \in \text{BMO}$, the singular integral operator $C[a]$ is defined by

$$C[a]\mu(x) = \frac{1}{\pi} \text{ p.v.} \int_\mathbf{R} \frac{\mu(y)}{(y-x) + i(A(y) - A(x))}\,dy,$$

where $A(x) = \int_0^x a(y)\,dy$. This is an expression of the Hilbert transform on the graph $\{x + iA(x); x \in \mathbf{R}\}$ by Calderón. He shows that $C[a]$ is $(L^1(\mathbf{R}), \text{weak } L^1(\mathbf{R}))$-bounded if $\|a\|_\infty$ is sufficiently small [**C1, C2**]. From this theorem and Corollary 5, the following theorem is deduced.

CALDERÓN-HAVIN-MARSHALL'S THEOREM ([**Ma**]). *If a compact set E on a rectifiable curve has a positive linear measure (i.e., $|E| > 0$), then $\gamma(E) > 0$.*

This is an answer to Denjoy's conjecture [**D, Gar**, p. 36] in the affirmative. Coifman, McIntosh, and Meyer [**CMM**] show the following: *For a real-valued function $a \in L^\infty(\mathbf{R})$, $C[a]$ is $(L^1(\mathbf{R}), \text{weak } L^1(\mathbf{R}))$-bounded.* This theorem is improved by David, Journé, and Semmes as the so-called $T(b)$ theorem (cf. [**Da1, DJ, DS, CJS**]). The recent developments on the singular integral operators of Calderón-Zygmund type are remarkable, and the detail is seen in [**Ch, Da3, Da4, DS, Me1, Me2**]. (These references are useful to see the relation between $\|\cdot\|_{p,p}$ and $\|\cdot\|_{1,w}$ also.) The following inequality is known as a norm estimate of $C[a]$ [**Mu1**, p. 53]: $\|C[a]\|_{1,w} \leq C_3\{1 + \sqrt{\|a\|_{BMO}}\}$ (C_3 *is an absolute constant*). This estimate is optimal [**Da2**]. From this inequality and Corollary 5, the following theorem is deduced.

THEOREM 6 ([**Mu1**], [**Mu2**]). *Let E be a compact set on a graph of length 1. Then $\gamma(E) \geq C_4 |\operatorname{pr} E|^{3/2}$, where $\operatorname{pr} E$ denotes the projection of E to $\mathbf{R}$ and C_4 is an absolute constant. The power $\frac{3}{2}$ is best possible.*

This theorem is a generalization of CHM's theorem. The exactness of the power $\frac{3}{2}$ is shown by using the cranks in the §5.

4. Arc-length variation

4.1. Functions Dc and D^2c. In this section, we present a variational approach to γ. We again confine ourselves to compact sets in $\mathscr{A}$. By (6), the variation of $\gamma(E)$, $E \in \mathscr{A}$ reduces to the Hadamard variation of the Szegö kernel function. The study of the variation of the Szegö kernel function is also seen in [**GS, HS, Sc1–Sc3, Sm, SS**]. In the classical theory, the local variational properties are mainly studied, and they are not applicable to γ directly. In order to apply the local variational properties to γ, it is necessary to study the Szegö kernel function from the viewpoint of **the continuation of the Hadamard variation for degenerate boundaries**. In this direction, two functions Dc and D^2c, which are defined later, play an important role. For a domain Ω, the pair $(K(z, \bar{\zeta}), L(z, \zeta))$ of the Szegö kernel function and the L-kernel function is defined uniquely; the condition "$K(\infty,\overline{\zeta}) = L(\infty,\zeta) = 0$" is removed if $\Omega \not\ni \infty$. For three mutually distinct numbers w, z, ζ in Ω, we set

$$Dc(z,\zeta;\Omega) = |L(z,\zeta)|^2 - |K(z,\overline{\zeta})|^2,$$

$$D^2c(w,z,\zeta;\Omega) = 2\Re\{DL(w,z,\zeta)\overline{L(z,\zeta)} - DK(w,z,\zeta)\overline{K(z,\overline{\zeta})}\},$$

where

$$DK(w,z,\zeta) = L(w,z)\overline{L(w,\zeta)} - \overline{K(w,\overline{z})}K(w,\overline{\zeta}),$$

$$DL(w,z,\zeta) = L(w,z)\overline{K(w,\overline{\zeta})} - \overline{K(w,\overline{z})}L(w,\zeta).$$

In this definition only, $K(\cdot,\overline{\infty}) = \overline{K(\infty,\cdot)}$ denotes $-g(\cdot;\Omega)$, and $L(\cdot,\infty) = -L(\infty,\cdot)$ denotes $-\phi(\cdot;\Omega)$. The functions Dc and D^2c are ε-coefficients of the following two variations, respectively, [**Hs, Sm, Mu5**]:

$$c(\zeta;\Omega \cup D(z,\varepsilon)) - c(\zeta;\Omega), \qquad Dc(z,\zeta;\Omega \cup D(w,\varepsilon)) - Dc(z,\zeta;\Omega).$$

These functions are metrics in a sense, and the following property is fundamental.

PROPOSITION 7. *The differential forms $Dc(z,\zeta)|dz||d\zeta|$, $D^2c(w,z,\zeta)|dw||dz||d\zeta|$ are conformally invariant, and the function Dc, D^2c are invariant for the permutations of variables.*

The Hadamard variation for degenerate boundaries is closely related to the differential equations of Löwner type. A parametrization $w_t: t \in [0,|E|] \to w_t \in E$ of $E \in \mathscr{A}$ is called an arc-length representation of E if w_t is continuous (w_t is right continuous at t_0 when $w_{t_0} \in E_{\text{en}}$) and $|E_t| = t$ ($0 \leq t \leq |E|$), where $E_t = \{w_s; 0 \leq s \leq t\}$. For an arc-length representation w_t, we write

$$Dc(w_t,z;E_t^c) = \lim_{u \downarrow t} Dc(w_u,z;E_t^c),$$

$$D^2c(w_t,z,\zeta;E_t^c) = \lim_{u \downarrow t} D^2c(w_u,z,\zeta;E_t^c).$$

(The limits exist.) Then the following theorem holds.

THEOREM 8 ([**Mu5**]).

$$\gamma(E) = \frac{1}{4}\int_0^{|E|} Dc(w_t,\infty;E_t^c)\,dt,$$

$$Dc(z,\infty;E_t^c) = Dc(z,\infty,E_s^c) + \frac{1}{4}\int_s^t D^2c(w_u,z,\infty;E_u^c)\,du \qquad (0 \leq s \leq t \leq |E|).$$

If $\Gamma \in \mathscr{A}$ is an arc, these formulae are expressed concretely. The Szegö kernel function and the L-function of a domain Γ_t^c ($0 \leq t \leq |\Gamma|$, $\Gamma_t = \{w_s; 0 \leq s \leq t\}$) are given by

$$\frac{\sqrt{q_t'(z)}\overline{\sqrt{q_t'(\zeta)}}}{q_t(z)\overline{q_t(\zeta)}-1}, \qquad \frac{\sqrt{q_t'(z)}\sqrt{q_t'(\zeta)}}{q_t(z)-q_t(\zeta)},$$

respectively, where q_t is a conformal mapping from Γ_t^c onto $\overline{\mathbf{D}}^c$ such that $q_t(\infty) = \infty$,

$q'_t(\infty) > 0$. Rewriting the second formula in Theorem 8 as a differential equation of q_t, we obtain

$$\text{(21)} \quad \frac{\partial q_t}{\partial t}(z) = \frac{1}{\gamma(\Gamma_t)} \frac{\partial \gamma}{\partial t}(\Gamma_t) q_t(z) \frac{1 + \kappa_t q_t(z)}{1 - \kappa_t q_t(z)} \quad (\kappa_t = \overline{q_t(w_t)}).$$

This is Löwner's D. E. [**A2**, p. 96] by an arc-length representation. (By the substitution $s = \log \gamma(\Gamma_t)$, we obtain the standard inverse form.) Thus, Theorem 8 is a generalization of Löwner's D. E. in terms of γ and plays an important role in the study of γ as Löwner's D.E. is basic in the theory of univalent functions. The following corollary is deduced from Theorem 8 and (21).

COROLLARY 9. *For an arc* $\Gamma \in \mathscr{A}$,

$$\gamma(\Gamma) = \frac{1}{4} \int_0^{|\Gamma|} \exp\left\{-\int_0^t \frac{1}{\gamma(\Gamma_s)} \frac{\partial \gamma}{\partial s}(\Gamma_s) \xi_s(w_t) \, ds\right\} dt,$$

where

$$\xi_s(w) = (\Im p_s(w))^2 / |1 - p_s(w)|^4 (\geq 0) \quad (p_s(w) = \kappa_s q_s(w)).$$

With the aid of this equality, we can estimate $\gamma(\Gamma)$ by the geometric properties of Γ.

4.2. Structure of D^2c. In order to handle a set function, it is important to know whether it is subadditive. Analytic capacity γ is a set function. Vitushkin [**V2**] shows that there are various applications to rational approximation if γ is semi-subadditive (i.e., $\gamma(A \cup B) \leq C_5\{\gamma(A) + \gamma(B)\}$, C_5 is an absolute constant). Davie [**Dav**] gives an equivalent test for the subadditivity of γ. At present, it is unknown whether γ is subadditive or not. As a positive step, the following theorem is known.

SUITA'S THEOREM ([**Su3**]). *If A and B are two continua, then*

$$\text{(22)} \quad \Gamma(A \cup B) \leq \gamma(A) + \gamma(B).$$

Suita deduces this theorem from Rengel's inequality [**T**, p. 415]; hence this theorem is related to the following Bieberbach inequality: *If $f(z)$ is a univalent function in* **D** *satisfying* $f(0) = \infty$ *and* $\lim_{z \to 0} |zf(z)| = 1$, *then* $|f'(z)| \geq |z|^{-2}(1 - |z|^2)$. Recall that $\text{Cap}(E) = \gamma(E)$ for a continuum E. Logarithmic capacity $\text{Cap}(\cdot)$ is not subadditive; however, Suita's Theorem shows that, for two continua A and B satisfying $A \cap B \neq \emptyset$, $\text{Cap}(A \cup B) \leq \text{Cap}(A) + \text{Cap}(B)$. His theorem suggests that γ is subadditive. Let $\mathscr{R}_n$ denote the totality of k-ply-connected domains ($k \leq n$), and let $\mathscr{K}_n$ denote the totality of compact sets consisting of continua less than or equal to n. Then Theorem 8 yields the following

PROPOSITION 10. *If the inequality $D^2c(\cdot,\cdot,\cdot,\cdot;\Omega) \leq 1$ is valid for all $\Omega \in \mathscr{R}_n$, then (22) holds as long as $A \cup B \in \mathscr{K}_n$, $A \in \mathscr{K}_m$, $B \in \mathscr{K}_l$ ($m + l \leq n + 1$). If the inequality $Dc(\cdot,\cdot;\Omega) \leq 1$ is valid for all $\Omega \in \mathscr{R}_n$, then (22) holds as long as $A \cup B \in \mathscr{K}_n$, $A \in \mathscr{K}_1$, $B \in \mathscr{K}_n$.*

For a Denjoy domain Ω (i.e., $\Omega^c \subset \mathbf{R}$), the pair (K, L) can be expressed by elementary functions [**B, Mu3, Mu5**]. Substituting the expression for (K, L) in $D^2 c$ and arranging the resulting equality, we obtain the following expression:

$$
\begin{aligned}
(23)\quad D^2 c(w, z, \zeta; \Omega) &= \frac{1}{4|M(w) M(z) M(\zeta)|} \\
&\times \Re \Bigg[\left(\frac{M(w) + M(z)}{w - z} \frac{\overline{M(w)} - \overline{M(\zeta)}}{\overline{w} - \overline{\zeta}} - \frac{\overline{M(w)} - \overline{M(z)}}{\overline{w} - \overline{z}} \frac{M(w) + M(\zeta)}{w - \zeta} \right) \\
&\quad \times \frac{\overline{M(z)} + \overline{M(\zeta)}}{\overline{z} - \overline{\zeta}} \\
&\quad - \left(\frac{M(w) + M(z)}{w - z} \frac{\overline{M(w)} + \overline{M(\zeta)}}{\overline{w} - \overline{\zeta}} - \frac{\overline{M(w)} - \overline{M(z)}}{\overline{w} - \overline{z}} \frac{M(w) - \overline{M(\zeta)}}{w - \overline{\zeta}} \right) \\
&\quad\quad \times \frac{\overline{M(z)} - \overline{M(\zeta)}}{\overline{z} - \overline{\zeta}} \Bigg],
\end{aligned}
$$

where

$$M(\xi) = \exp\left\{ \frac{1}{2} \int_{\Omega^c} \frac{dt}{t - \xi} \right\} \qquad (\xi = w, z, \zeta).$$

An n-ply-connected domain ($n = 1, 2, 3$) is conformally equivalent to a Denjoy domain, and hence, (23) is applicable to all simply, doubly, and triply connected domains. Computing (23) practically, we obtain the following

THEOREM 11 ([**Mu5**]). *If $\Omega \in \mathscr{R}_2$, then $D^2 c(\cdot, \cdot, \cdot; \Omega) \leq 0$.*

From this theorem and Proposition 10, Suita's Theorem is deduced. The computation of (23) for a triply connected domain is also possible. The computation of (23)

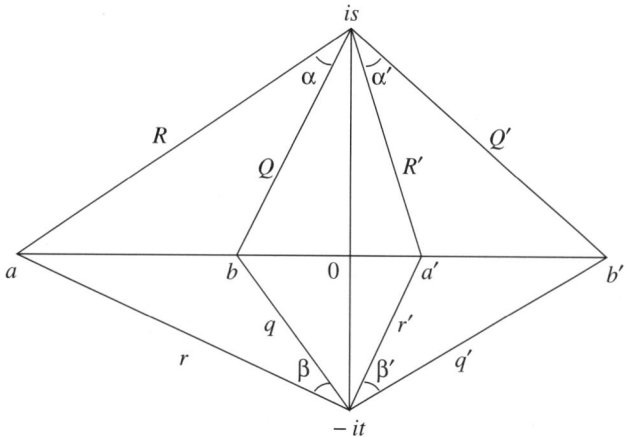

$(a < b < 0 < a' < b' ; R, R', Q, Q', r, r', q, q', \alpha, \alpha', \beta, \beta', s, t > 0)$

FIGURE 1

for a doubly connected Denjoy domain $\{[a,b] \cup [a',b']\}^c$ shows that the inequality $D^2c \leq 0$ reduces to the following inequality of Möbius type in elementary geometry (see Figure 1):

$$\frac{QQ'rr'}{RR'qq'} + \frac{RR'qq'}{QQ'rr'} \leq 2\cos(\alpha+\alpha')\cos(\beta+\beta') + \frac{s^2+t^2}{st}\sin(\alpha+\alpha')\sin(\beta+\beta').$$

Theorem 11 suggests that the structure of D^2c is better than we have expected. It seems important to study D^2c even for Denjoy domains.

5. Cranks

5.1. Integral geometry. Each finitely connected domain is conformally equivalent to a parallel-slit domain i.e., a domain whose complement is an element of $\mathscr{G}$. Here $\mathscr{G}$ denotes the totality of finite unions of segments parallel to $\mathbf{R}$. Thus, it is reasonable, as a method of studying γ, to restrict our attention to elements in $\mathscr{G}$. Fractal sets [M] appear in various extremum problems about γ as a set function on $\mathscr{G}$. The configuration of a fractal element in $\mathscr{G}$ is like a crank. From this fact, such an element is called a crank. Cranks are not artificial sets but solutions of various extremum problems. The extreme of cranks is applied to the construction of examples. In this section, we define a class $\mathscr{F}$ of cranks which is used to compare γ with various set functions (for example, Theorems 6 and 12), and we note Besicovitch's integral geometry which is closely related to cranks. Fix $\rho_0 > 0$. For a natural number m, we set

$$\Gamma([0,1],m) = \left\{ \bigcup_{j=0}^{[(m-1)/2]} \left(\frac{2j}{m} + \left[0, \frac{1}{m}\right]\right)\right\} \cup \left\{ \bigcup_{j=0}^{[(m-2)/2]} \left(\frac{2j+1}{m} + \frac{i\rho_0}{m} + \left[0, \frac{1}{m}\right]\right)\right\},$$

and, for $E = \bigcup_{k=1}^n I_k \in \mathscr{G}$ (each I_k is a segment) and natural numbers $m_1, m_2, \ldots, m_n$, we set

$$\Gamma(E, m_1, \ldots, m_n) = \bigcup_{k=1}^n (a_k + |I_k|\Gamma([0,1], m_k)),$$

where a_k is the left endpoint of I_k. Define a sequence $\mathscr{F}_l$ ($l \geq 0$) of families of cranks inductively by $\mathscr{F}_0 = \{[0,1]\}$,

$$\mathscr{F}_l = \left\{ \Gamma(E, m_1, \ldots, m_n); E = \bigcup_{k=1}^n I_k \in \mathscr{F}_{l-1}, m_1, \ldots, m_n \geq 1 \right\} \quad (l \geq 1),$$

and set $\mathscr{F} = \bigcup_{l=0}^\infty \mathscr{F}_l$. The family $\mathscr{F}$ of cranks has the following properties. The total length of each element of $\mathscr{F}$ is equal to 1, and the projection to $\mathbf{R}$ is the interval $[0,1]$. There exists a constant C_6 depending only on ρ_0 such that

(24) $\qquad |E \cap D(z,r)| \geq C_6 r \qquad (0 < r < 1, z \in E, E \in \mathscr{F})$

(i.e., each element of $\mathscr{F}$ is uniformly thick in the sense of Ahlfors). Let $\mathscr{A}_E$ denote the arc obtained by adding $E \in \mathscr{F}$ and a finite number of segments perpendicular to the x-axis. Then

$$\rho_0 l/2 + 1 \leq |\mathscr{A}_E| \leq \rho_0 l + 1 \qquad (E \in \mathscr{F}_l,\ l \geq 1).$$

The extremal properties of cranks in $\mathscr{F}$ are based on these properties (cf. the traveling salesman problem [**BJ, J1, J2**]). A set

$$CR_l = \bigcup_{k=1}^{2^l} \left\{ [(k-1)2^{-l}, k2^{-l}] - i \sum_{j=1}^{l} (-1)^j \left(\delta_{k,j} - \frac{1}{2} \right) 2^{-j} \right\}$$

is a crank in $\mathscr{F}$ (see Figure 2), where $\{\delta_{k,j}\}$ denote the coefficients of the dyadic expansion of $(k-1)2^{-l}$ i.e.,

$$(k-1)2^{-l} = \sum_{j=1}^{l} \delta_{k,j} 2^{-j}, \qquad \delta_{k,j} = 0 \text{ or } 1.$$

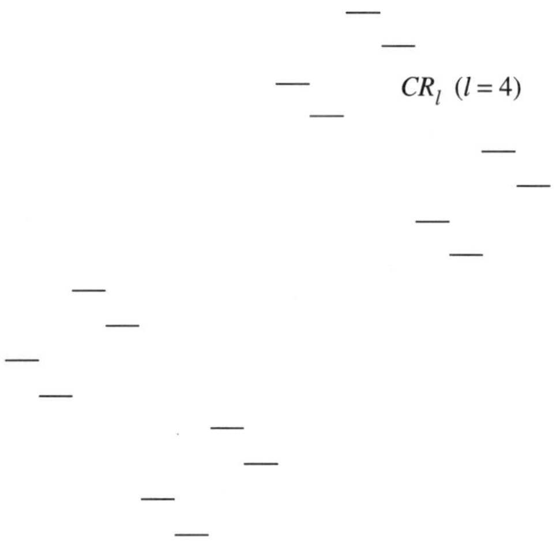

CR_l ($l = 4$)

FIGURE 2

Cranks appear in various areas, and it is well-known that Besicovitch's integral geometry is related to cranks. Let $\mathscr{B}$ denote the totality of Borel sets having finite 1-dimensional Hausdorff measure $|\cdot|$. The notation a.e. means that a property holds except for a set of zero 1-dimensional Hausdorff measure. The lower density of $E \in \mathscr{B}$ at $z \in \mathbf{C}$ is defined by

$$\underline{d}(z, E) = \liminf_{r \to 0} \frac{|E \cap D(z,r)|}{2r},$$

and the upper density $\overline{d}(z, E)$ of E at z is defined with "lim inf" replaced by "lim sup". The equality $\overline{d}(z, E) = 1$ holds a.e. on E. A set $E \in \mathscr{B}$ is regular if $\underline{d}(z, E) = 1$ a.e. on E, and E is irregular if $\underline{d}(z, E) < 1$ a.e. on E. The straight line $x \sin \theta - y \cos \theta = 0$

$(-\pi/2 < \theta \leq \pi/2)$ passing through the origin is denoted by $\mathbf{L}_\theta$. The Favard length (i.e., the Buffon needle probability) of E is defined by

$$Bu(E) = \int_{-\pi/2}^{\pi/2} |\operatorname{pr}_\theta E| \, d\theta,$$

where $\operatorname{pr}_\theta E$ denotes the projection of E to $\mathbf{L}_\theta$. We define the straight line $\mathbf{L}_\perp(\theta, r)$ ($r \in \mathbf{R}, -\pi/2 < \theta \leq \pi/2$) by $x\cos\theta + y\sin\theta = r$, and we define a function $n_E(\theta, r)$ by

$$n_E(\theta, r) = 1 \quad (E \cap \mathbf{L}_\perp(\theta, r) \neq \varnothing),$$
$$n_E(\theta, r) = 0 \quad (E \cap \mathbf{L}_\perp(\theta, r) = \varnothing).$$

Then

$$\operatorname{Bu}(E) = \int_0^\infty \left\{ \int_{-\pi/2}^{\pi/2} n_E(\theta, r) \, d\theta \right\} dr.$$

(From this fact, $\operatorname{Bu}(E)$ is also called the Buffon needle probability.)

BESICOVITCH'S THEOREM ([F]).
(25) A set $E \in \mathcal{B}$ is decomposable into a regular set and an irregular set (*Besicovitch's decomposition*).
(26) A set $E \in \mathcal{B}$ is regular if and only if there exist countable rectifiable curves $\{\gamma_k\}$ such that $|E - \bigcup \gamma_k| = 0$.
(27) A set $E \in \mathcal{B}$ is irregular if and only if $\operatorname{Bu}(E) = 0$.
(28) If $E \in \mathcal{B}$ is irregular, then $\underline{d}(z, E) < 3/4$ a.e. on E.
(29) If $E \in \mathcal{B}$ is irregular, then $|\operatorname{pr}_\theta E| > 0$ holds for at most one direction θ.

The meaning of the regular part E_0 in Besicovitch's decomposition (25) is "$\underline{d}(z, E_0) = 1$ a.e. on E_0" and is not "a.e. on E". Roughly speaking, (28) shows that the lower density of an irregular set does not take the value in an interval $(3/4, 1)$. (As the best possible value, $1/2$ is conjectured.) Property (29) yields that $|E_0| > 0$, if $|\operatorname{pr}_\bullet E_0| > 0$ holds for two different directions.

The configuration of an irregular set is like a limit set of cranks, and cranks in $\mathcal{F}$ yield irregular sets of graph type. Suppose, for example, that a sequence $\{E_l\}$, $E_l \in \mathcal{F}$ satisfies the following condition: Each E_l is expressed as $E_l = \Gamma(E_{l-1}, \ldots)$ so that there is no common component between E_{l-1} and E_l. Then the limit set E_∞ of this sequence is an irregular set satisfying (24) and $\operatorname{pr} E_\infty = [0, 1]$. The solution of the following extremum problem is also a fractal set nearly equal to an element in $\mathcal{F}$:

$$\sigma_n = \inf\{\operatorname{Bu}(E); E \in \mathcal{G}, |\operatorname{pr} E| = 1, E \text{ consists of } n \text{ components}\}.$$

A simple calculation shows that

$$\sigma_2 = \operatorname{Bu}\left(\left[0, \frac{1}{2}\right] \cup \left(\left[\frac{1}{2}, 1\right] + i\rho_0\right)\right)$$

$$= 2 - \left\{\rho_0 + \sqrt{\rho_0^2 + 1} - 2\sqrt{\rho_0^2 + \left(\frac{1}{4}\right)}\right\} = 1.79 \cdots$$

$$\left(\rho_0 = \frac{1}{2}\sqrt{2\sqrt{3}\cos(5\pi/18) - 1} = 0.553 \cdots\right).$$

5.2. Vitushkin-Garnett's sets.
A compact set having no interior point satisfies $\gamma(E) \leq |E|/\pi$. (The author conjectures that $\gamma(E) \leq |E|/4$.) The following assertion shows that the inequality in the inverse direction does not hold in general.

VITUSHKIN-GARNETT'S THEOREM ([**Gar**], [**V1**]). *There exists a compact set E such that $|E| > 0$ and $\gamma(E) = 0$.*

This theorem was first shown by Vitushkin [**V1**], and a simplified example was given by Garnett [**Gar,** p. 87] later. Garnett's example $Q_\infty = Q_\infty(1/4)$ is constructed as follows. Take the unit square $Q_0(\alpha) = [0,1] \times [0,1]$ $(0 < \alpha < 1/2)$. Let $Q_l(\alpha)$ be the union of 4^l squares of area α^{2l} in the 4^l corners of 4^{l-1} components of $Q_{l-1}(\alpha)$ $(l = 1, 2, \ldots)$, and put $Q_\infty(\alpha) = \bigcap_{l=0}^\infty Q_l(\alpha)$. The computation of the Hausdorff dimension of $Q_\infty(\alpha)$ shows that $\gamma(Q_\infty(\alpha)) = |Q_\infty(\alpha)| = 0$ $(0 < \alpha < 1/4)$ and $\gamma(Q_\infty(\alpha)) > 0$ $(1/4 < \alpha < 1/2)$. Thus, the case of $\alpha = 1/4$ is critical. Evidently, $|Q_\infty| = 1\sqrt{2}$. For the proof of $\gamma(Q_\infty) = 0$, the following equality is important: $\text{pr}_{\theta_0} Q_l(1/4) = e^{i\theta_0}[0, 3/\sqrt{5}]$ ($\theta_0 = \arctan 1/2$). From this equality, we can regard $Q_l(1/4)$ as a crank. (Compare the configurations of CR_4 and $Q_2(1/4)$.) There are several methods of the proof of $\gamma(Q_\infty) = 0$, and each proof has its own merit.

GARNETT-MATTILA'S METHOD ([**Gar,** p. 87], [**Mat1**]). This is a method to deduce a contradiction "$f_{Q_\infty} \notin H^\infty(Q_\infty^c)$" from the assumption that the Ahlfors function f_{Q_∞} of Q_∞ is not 0. This method is related to Besicovitch's integral geometry and has some applications.

JONES'S METHOD ([**J1,** p. 51]). Let $\text{BMO}(\Omega_l)$ denote the BMO-space of analytic functions f in a domain $\Omega_l = Q_l(1/4)^c$ with norm

$$\sup_{z \in \Omega_l} \left\{ \int\!\!\int_{\Omega_l} G_l(z,\zeta) |f'(\zeta)|^2 \, dx \, dy \right\}^{1/2} \qquad (\zeta = x + iy),$$

where $G_l(z, \zeta)$ is the Green's function of Ω_l. Evidently, $H^\infty(\Omega_l) \subset \text{BMO}(\Omega_l)$. Jones deduces the property $\bigcap_{l=0}^\infty \text{BMO}(\Omega_l) = \{0\}$ from an estimate of the Green's function. This implies that Q_∞ is a BMO-null set (which is a stronger assertion than $\gamma(Q_\infty) = 0$).

DYPOLE METHOD ([**Mu1,** p. 81]). This is a method to construct a function $\psi_l \in H^1(\Omega_l)$, $|\psi_l(\infty)| = 1$ with a small H^1-norm; notice that $\gamma(Q_\infty) \leq \|\psi\|_{H^1}/(2\pi)$ ($\psi \in H^1(\Omega_l)$, $|\psi(\infty)| = 1$). For the construction of ψ_l, we use Rankin's combined vortex

$$\phi(z) = \phi(z; \zeta, \zeta', w, w')$$
$$= ie^{-i\alpha} \frac{(z-w)(z-w')}{(z-\zeta)(z-\zeta')} \qquad \left(\zeta, \zeta', w, w' \in \mathbf{C}, \ \alpha = \arg \frac{ww' - \zeta\zeta'}{\zeta - \zeta'} \right),$$

where w, w' are two points on the segment $\overline{\zeta\zeta'}$ satisfying $w + w' = \zeta + \zeta'$. Rankin's combined vortex ϕ satisfies the property

(30) $\qquad \phi(z) = 1 + O(|z|^{-2}) \ (z \to \infty), \qquad |\phi(z)| < 1 \ (z \in U),$

where U is a strip containing $\overline{ww'}$ with width $|w - w'|$. Taking (30) into consideration, we choose vortex centers ζ_k, ζ_k' and stagnation points w_k, w_k' $(1 \leq k \leq m)$ suitably. Then $\psi_l = \prod_{k=1}^m \phi(\cdot; \zeta_k, \zeta_k', w_k, w_k')$ is the required function. This method is related to the theory of stable vortices, and suggests that the sets of Vitushkin-Garnett type have some fluid dynamical meaning.

5.3. Galton-Watson process. In this section, we show that the computation of $Bu(\cdot)$ for cranks is closely related to the Galton-Watson process. To do this, we define the Galton-Watson process from a geometric property of cranks. The unit interval $K_0 = [0, 1]$ is regarded as a probability space with respect to $|\cdot|$. Let $\{\varepsilon_l\}$ denote Rademacher's sequence on K_0, i.e., $\{\varepsilon_l\}$ is independent and

$$\Pr(\varepsilon_l = 1) = \Pr(\varepsilon_l = -1) = 1/2 \qquad (l \geq 1);$$

$\Pr(X)$ denotes the probability that the event X occurs. The random walk $\{S_l; l \geq 0\}$ is defined by

$$S_l = \sum_{k=1}^{l} \varepsilon_k \quad (l \geq 1), \qquad S_0 = 0.$$

The Galton-Watson process $\{y_l\}$ is defined by $y_0 = 1$,

$$y_l(x) = y_{l-1}(x) + S_{y_{l-1}(x)}(x) \qquad (x \in K_0, l \geq 1).$$

The following example is frequently quoted as an explanation of this process: In a game that one gets (or loses) a coin with probability $1/2$ each second, a man who has n coins counts the number $n + S_n$ of his coins after n seconds. The generating function

$$P_l(t) = \sum_{k=0}^{\infty} \Pr(y_l = k) t^k$$

of y_l satisfies the following inductive equation:

$$P_l(t) = P_{l-1}\left(\frac{1+t^2}{2}\right) \quad (l \geq 1), \qquad P_0(t) = t.$$

From this equation, we can deduce

(31) $$C_7/l \leq \Pr(y_l > 0) \leq C_8/l \qquad (l \geq 1),$$

where C_7 and C_8 are absolute constants. This Galton-Watson process is defined from cranks as follows (see Figure 3).

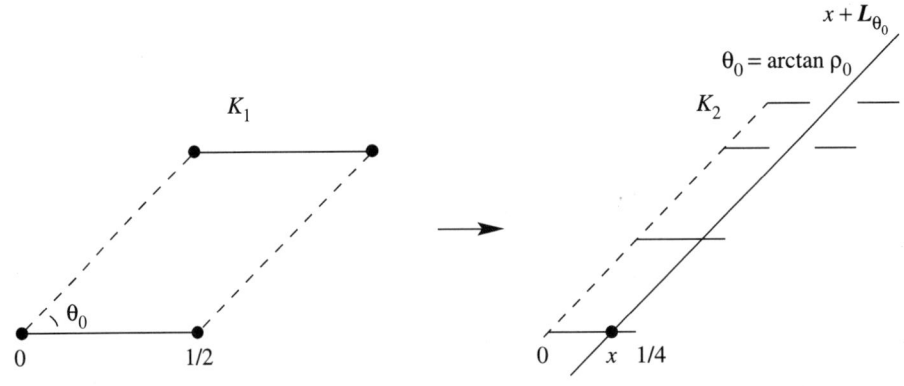

FIGURE 3

Let $y(x; E, \theta)$ denote the number of the intersections between $E \in \mathscr{F}$ and the straight line $x + \mathbf{L}_\theta$ ($0 \leq x \leq 1$, $|\theta| \leq \pi/2$). For an even number $m \geq 2$ and a segment I parallel to $\mathbf{R}$, we set

$$\widetilde{I} = I \cup I', \qquad \widetilde{\Gamma}(I, m) = \Gamma(I, m) \cup \Gamma(I', 2m),$$

where $I' = I + (1 + i\rho_0)|I|$. A set $\widetilde{\Gamma}(I, m)$ is expressed as a union of doubles (segment)$^\sim$. The sequence $\{K_l\}$, $K_l \in \mathscr{F}_l$ of cranks is defined inductively as follows. Set $K_1 = [0, 1/2]^\sim$. Once $K_l = \bigcup_{k=1}^{p_l} \widetilde{I}_k^{(l)}$ (each $I_k^{(l)}$ is a segment) has been determined, we set

$$K_{l+1} = \bigcup_{k=1}^{p_l} \widetilde{\Gamma}(I_k^{(l)}, 2^{1-l}|I_k^{(l)}|^{-1}).$$

Then $\{y(\cdot; K_l, \theta_0)\}$ ($\theta_0 = \arctan \rho_0$) is the required Galton-Watson process.

Using this Galton-Watson process, we can estimate $\Pr(y(\cdot; E, \theta) > 0)$. Combine with Theorem 6, the following theorem yields (cf. [**Ma, Mat2, Mu1**]).

THEOREM 12 ([**JM**]). *There exists a compact set E_0 such that $\gamma(E_0) > 0$ and $\mathrm{Bu}(E_0) = 0$.*

Recall that $\mathrm{Bu}(\cdot) \leq 2|\cdot|$. The following inequalities are conjectured: $\mathrm{Bu}(\cdot) \leq 8\gamma(\cdot) \leq 2|\cdot|$.

6. Aerodynamics

6.1. Mathematical formulation. In this section, we give an application of our method to aerodynamics. In the discussion of 2-dimensional flow of the high-speed air, the theory of inviscid fluids fits actually phenomena. The hodograph method and the M^2 expansion method enable us to regard subsonic flow as a perturbation of incompressible flow. Thus, the theory of incompressible flow of inviscid fluid is indispensable even nowadays. A problem on 2-dimensional ideal fluid is mathematically nothing less than a boundary value problem of analytic functions. Two Milne-Thomson's books [**Mi1, Mi2**] are classical compendia and Betz's book [**Bet**] on conformal mappings are written from the viewpoint of hydrodynamics. Throughout this chapter, $E = \bigcup_{k=1}^n \Gamma_k \in \mathscr{A}$, $\Omega = E^c$. For $-\pi < \alpha \leq \pi$ and an analytic function f, $f(\infty) = 0$ in Ω, the function $\mathbf{Z} = \overline{e^{i\alpha} - if}$ is regarded as a velocity field outside (the section of) an obstacle E. Then the aerodynamics force (i.e., the lift) induced by $\mathbf{Z}$ (with respect to the density 1) is given by

$$\mathscr{L}(E, \mathbf{Z}) = -\overline{\frac{i}{2} \int_{\partial \Omega} \{e^{i\alpha} - if\}^2 \, dz} = 2\pi i e^{-i\alpha} \overline{f'(\infty)}.$$

The velocity field $\mathbf{Z}$ is determined uniquely by the following three conditions:

(32) Flow induced by $\mathbf{Z}$ coincides with the configuration of E on the boundary $\partial \Omega$.
(33) f is expressed as $f = \mathscr{C}\mu$ for some $\mu \in L^1(E)$.
(34) $\mathbf{Z}$ satisfies the Kutta-Joukowski condition.

Condition (32) implies that a fluid flows along an obstacle. Condition (33) means that a velocity field is induced by a set of vortex filaments. Choosing an endpoint of each Γ_k, we fix n endpoints of E. These n endpoints are called the trailing edges, and the other n endpoints are called the leading edges. The Kutta-Joukowski condition (34) is an assumption that a boundary value of $\mathbf{Z}$ exists at each trailing edge. The propriety of this condition is based on the boundary-layer theory and the experimentation [**Sch**]. These conditions are ideals; actual flow is unstable and Kármán's vortex sheets appear behind an obstacle. To rewrite the above three conditions, we introduce some notation. The singular integral operators $\mathscr{H}_r$, $\mathscr{H}_i$ are defined from the kernels $\frac{1}{\pi}\Re 1/(\zeta-z)$, $\frac{1}{\pi}\Im 1/(\zeta-z)$, respectively. Evidently, $\mathscr{H} = \mathscr{H}_r + i\mathscr{H}_i$. A function in E^{in} is defined by $h_\alpha = \rho_r \sin\alpha + \rho_i \cos\alpha$, where $\rho_r = \Re\rho$, $\rho_i = \Im\rho$, $\rho = dz_+/|dz_+|$. The operator $\mathscr{H}_0$ on $L^1(E)$ is defined by $\mathscr{H}_0\mu = \rho_r\mathscr{H}_r\mu - \rho_i\mathscr{H}_i\mu$. Then the above three conditions are equivalent to the following two conditions:

(35) $\mathscr{H}_0\mu = h_\alpha$ $(\mu \in L^1(E))$.

(36) μ is real-valued, continuous in E^{in} and takes 0 at each trailing edge.

The relation with complex velocity potentials is as follows. The complex velocity potential of a real-valued function $\mu \in L^1(E)$ is defined by

$$U\mu(z) = \frac{1}{\pi}\int_E \log\frac{1}{\zeta-z}\mu(\zeta)|d\zeta|.$$

This is a multivalued analytic function, and its real part

$$U_r\mu(z) = \frac{1}{\pi}\int_E \log\frac{1}{|\zeta-z|}\mu(\zeta)|d\zeta|$$

is single-valued. For a smooth function H on E, there exists uniquely a pair (μ_E, c_E) of a real-valued function $\mu_E \in L^1(E)$ and $c_E \in \mathbf{R}$ such that $U_r\mu_H = H + c_H$ (semibalyage principle). Let H_α be the function on E vanishing at each trailing edge such that $\partial H_\alpha/\partial s_z = h_\alpha$, where $\partial/\partial s_z$ denotes the derivative along E satisfying $\partial z_+/\partial s_z = \rho$. Evidently,

$$\mathscr{H}_0\mu_{H_\alpha} = \partial(U_r\mu_{H_\alpha})/\partial s_z = h_\alpha,$$

i.e., μ_{H_α} is a solution of (35). We define H_k by $H_k = 1$ on Γ_k, $H_k = 0$ on $E - \Gamma_k$. Then the space of solutions of (35) is expressed as

$$\mu_\alpha^* = \mu_\alpha^*[d_1,\ldots,d_n] = \mu_{H_\alpha} + \sum_{k=1}^n d_k\mu_{H_k} \qquad (d_1,\ldots,d_n \in \mathbf{R})$$

(the sum of a special solution and general solutions of a homogeneous equation which form an n-dimensional real vector space). The imaginary part of the derivative of the complex velocity potential $e^{i\alpha} - iU\mu_\alpha^*$ along E is 0. Thus, the imaginary part of the complex velocity potential on E is a constant on each component (cf. [**Ko**]). In order to solve (35) and (36), it is sufficient, in principle, to determine $d_1,\ldots,d_n \in \mathbf{R}$ so that $\mu_\alpha^*[d_1,\ldots,d_n]$ satisfies (36).

6.2. Algorithm for the computation and the interaction coefficient. For the computation of the lift, it is not suitable to use the Green's function based on Dirichlet's principle. In this section, we give a practical method of the construction of the solution of (35) and (36). We define an operator $\mathscr{H}_n$ on $L^1(E)$ by $\mathscr{H}_n\mu = \rho_i\mathscr{H}_r\mu + \rho_r H_i\mu$. This is a compact operator on $L^1(E)$ and satisfies $\mathscr{H}_0\mu = \rho\mathscr{H}\mu - i\mathscr{H}_n\mu$ $(\rho = dz_+/|dz_+|)$. This fact and (19) in Theorem 3 yield the following

THEOREM 13. *There exists uniquely* $\mu_\alpha \in L^1(E)$ *satisfying* (35) *and* (35).

We say that $E \in \mathscr{G}$ is RKJ-conditioned if each right endpoint is chosen as a trailing edge. (The RKJ-condition is the abbreviation of the Kutta-Joukowski condition for right endpoints.) When $E \in \mathscr{G}$ is RKJ-conditioned, (35) is simplified in the form $\mathscr{H}_r\mu = \sin\alpha$. The practical algorithm for the construction of μ_α is as follows.

THEOREM 14 ([**Mu7**]). *Suppose that $E \in \mathscr{G}$ is RKJ-conditioned. Choose a positive number R_0^* so that $R_0^* > \|\mathscr{H}_r\|_{2,2}$. For $0 < c < 1$, we define a sequence $\{v_l^*\} \subset L^2(E)$ inductively by $v_0^* = 0$,*

$$R_0^* v_l^* - \mathscr{H}_r v_l^* = (R_0^* - c) v_{l-1}^* - \sin\alpha \qquad (l \geq 1). \tag{37}$$

Then the limit $v_c = \lim_{l\to\infty} v_l^* \;(\in L^2(E))$ exists and $\mu_\alpha = \lim_{c\to 0} v_c \;(\in L^1(E))$.

Notice that the computation of the inverse operator of the lift-hand side of (37) is possible. For a general $E \in \mathscr{A}$, μ_α is constructed as a perturbation of the solution for an RKJ-conditioned element in $\mathscr{G}$.

The function $\mathbf{Z}_\alpha = e^{i\alpha} - i\mathscr{C}\mu_\alpha$ is called the velocity field induced by $(E, e^{-i\alpha})$. The lift and the interaction induced by $(E, e^{-i\alpha})$ are defined, respectively, by

$$\mathscr{L}_\alpha(E) = \mathscr{L}(E, \mathbf{Z}_\alpha), \qquad \sigma_\alpha(E) = \mathscr{L}_\alpha(E)/\sum_{k=1}^n \mathscr{L}_\alpha(\Gamma_k).$$

Here, in the definition of $\mathscr{L}_\alpha(\Gamma_k)$, it is required that the endpoint of Γ_k which is used as one of the trailing edges of E is adopted as the trailing edge. The lift coefficient and the interaction coefficient induced by E are defined by

$$\mathscr{L}(E) = \max_{|\alpha| \leq \pi} |\mathscr{L}_\alpha(E)|, \qquad \sigma(E) = \max_{|\alpha| \leq \pi} |\sigma_\alpha(E)|,$$

respectively. If $\sigma(E)$ is small, then E behaves like a small obstacle in flow even if $|E|$ is large. Concerning the interaction for several obstacles, the following theorem holds.

THEOREM 15 ([**Mu7**]). *For an RKJ-conditioned $E \in \mathscr{G}$, $E \neq \varnothing$, the inequalities $\mathscr{L}(E) > 0$ and $\sigma(E) \leq 1$ hold. For $E \subset \mathbf{R}$, the equality $\sigma(E) = 1$ holds. For any $\varepsilon > 0$, there exists an RKJ-conditioned $E_\varepsilon \in \mathscr{G}$ such that the projection to $\mathbf{R}$ is the unit interval $[0, 1]$ and $\sigma(E_\varepsilon) \leq \varepsilon$.*

When one looks up to E_ε from the ground, this appears as one connected obstacle with length 1, however, E_ε behaves like a small obstacle in flow. For the construction of E_ε, cranks are used. Choosing a large constant C_9, we put $E_\varepsilon = CR_{l_\varepsilon}$ ($l_\varepsilon = [e^{C_9/\varepsilon}]$). Then E_ε is the required set. (See the figure in §5.1.) In the proof, the inequality $\mathscr{L}(E) \leq 4\pi\gamma(E)$ ($E \in \mathscr{G}$) plays an important role.

6.3. Theory of two aeroplanes. The theory of two aeroplanes is classical and it is mathematically nothing less than the theory of elliptic functions [**Fe, Gk, Mu4**]. The first step in this theory is to obtain a concrete expression of a conformal mapping from a canonical domain onto a given doubly connected domain. For an obstacle $E = (w + [0, r]) \cup (z + [0, r])$ ($w, z \in \mathbf{C}$, $r > 0$) consisting of two segments parallel to $\mathbf{R}$, we form the following Schwarz-Christoffel transform:

$$f(\zeta) = \int_0^\zeta \frac{s^2 - m_k^2}{\sqrt{s-1}\sqrt{s+1}\sqrt{ks-1}\sqrt{ks+1}} ds - it\zeta \tag{38}$$
$$\left(km_k = \sqrt{E(k')/K(k')},\ 0 < k < 1,\ t > 0\right),$$

where K, E denote the complete elliptic integrals of the first kind and of the second kind, respectively, and $k' = \sqrt{1-k^2}$ (complementary modulus). Then we can choose k and t so that f is a conformal mapping from J^c ($J = [-1/k, -1] \cup [1, 1/k]$) onto iE^c. (The Schwarz-Christoffel transform (38) is classical, and the author learned (38) from [**Sas**].) In the case of two segments with different length, the analogous method also works. In the case of two radial slits, Weierstraß's elliptic functions [**H**] are useful. Analytic capacity is expressed by the lift-coefficient as follows.

THEOREM 16 ([**Mu4**]). *Let* $\Gamma(z) = [-1/2, 1/2] \cup (z + [-1/2, 1/2])$ *and suppose that it is RKJ-conditioned. Then*

$$\gamma(\Gamma(z)) = \frac{1}{2} + \frac{\Im z}{2} \int_{\lambda(z)} \left\{ \frac{4\pi\gamma(\Gamma(\zeta))}{\mathscr{L}(\Gamma(\zeta))} - 1 \right\} \frac{d(\Im\zeta)}{(\Im\zeta)^2} \quad (z \in \mathbf{C}_{++}),$$

where $\mathbf{C}_{++}$ *denotes the first quadrant and* $\lambda(z)$ *is the arc starting from z and ending at a positive number such that the modulus of* $\Gamma(\zeta)^c$ [**SO**, p. 272] *is a constant on* $\lambda(z)$ *(as a function of ζ).*

From this theorem, the following equality is deduced:

$$\min_{x,y \in \mathbf{R}} \frac{\gamma(\Gamma(x+iy))}{\gamma(\Gamma(x))} = \min_{y>0} \frac{\gamma(\Gamma(1+iy))}{\gamma(\Gamma(1))} (= 0.9 \cdots).$$

This equality gives a reason why cranks play an important role in the study of sets of Vitushkin-Garnett type.

References

[A1]   L. Ahlfors, *Bounded analytic functions*, Duke Math. J. **14** (1947), 1–11.
[A2]   _____, *Conformal invariants: Topics in geometric function theory*, McGraw-Hill, New York, 1973.
[AB]   L. Ahlfors and A. Beurling, *Conformal invariants and function-theoretic null-sets*, Acta Math. **83** (1950), 101–129.
[B]   W. Baker II, *Kernel functions on domains with hyperelliptic double*, Trans. Amer. Math. Soc. **231** (1977), 339–347.
[Be]   S. Bergman, *The kernel function and conformal mapping*, Math. Surveys, vol. 5, Amer. Math. Soc., Providence, RI, 1950.
[Bet] A. Betz, *Konforme Abbildung*, 2 Auflage, Springer-Verlag, Berlin, 1964.
[BJ]   C. Bishop and P. Jones, *Harmonic measure and arclength*, Ann. of Math. **132** (1990), 511–547.
[C1]   A. P. Calderón, *Cauchy integrals on Lipschitz curves and related operators*, Proc. Nat. Acad. Sci. U.S.A. **74** (1977), 1324–1327.
[C2]   _____, *Commutators, singular integrals on Lipschitz curves and applications*, Proc. Internat. Congress Math. (Helsinki, 1978), Amer. Math. Soc., Providence, RI, 1980, pp. 85–96.
[Ch]   F. M. Christ, *Lectures on singular integral operators*, CBMS Regional Conf. Ser. in Math., no. 77, Amer. Math. Soc., Providence, RI, 1990.
[CJS] R. R. Coifman, P. W. Jones, and S. Semmes, *Two elementary proofs of L^2 boundedness of Cauchy integrals on Lipschitz curves*, J. Amer. Math. Soc. **2** (1989), 553–564.
[CMM] R. R. Coifman, A. McIntosh, and Y. Meyer, *L'intégrale de Cauchy définit un opérateur borné sur L^2 pour les courbes lispchitziennes*, Ann. of Math. **116** (1982), 361–388.
[D]   A. Denjoy, *Sur la continuité des fonctions analytiques singulières*, Bull. Soc. Math. France **60** (1932), 27–105.
[Da1] G. David, *Opérateurs intégraux singuliers sur certaines courbes du plan complexe*, Ann. Sci. École Norm. Sup. **17** (1984), 157–189.
[Da2] _____, *Noyau de Cauchy et opérateurs de Calderón-Zygmund*, Thèse d'Etat, Orsay (1986).
[Da3] _____, *Opérateurs de Calderón-Zygmund*, Proc. Internat. Congress Math (Berkeley, 1986), Amer. Math. Soc., Providence, RI, 1987, pp. 890–899.
[Da4] _____, *Wavelets, Calderón-Zygmund operators, and singular integral operators on surfaces*, Lecture Notes in Math., vol. 1465, Springer-Verlag, Berlin, 1991.

[Dav]   A. M. Davie, *Analytic capacity and approximation problems*, Trans. Amer. Math. Soc. **171** (1972), 409–444.
[DØ]   A. M. Davie and B. Øksendal, *Analytic capacity and differentiability properties of finely harmonic functions*, Acta Math. **149** (1982), 127–152.
[DJ]   G. David and J. L. Journé, *A boundedness criterion for generalized Calderón-Zygmund operators*, Ann. of Math. **120** (1984), 371–397.
[DS]   G. David and S. Semmes, *Singular integrals and rectifiable sets in $\mathbf{R}^n$: Au-delà des graphes lipschitziens*, Astérisque **193**, Soc. Math. France, 1991.
[F]   K. J. Falconer, *The geometry of fractal sets*, Cambridge Univ. Press, Cambridge, 1985.
[Fa]   J. D. Fay, *Theta functions on Riemann surfaces*, Lecture Notes in Math., vol. 352, Springer-Verlag, Berlin, 1973.
[Fe]   C. Ferrari, *Sulla transformazione conforme di due cerchi in due profili alari*, Mem. Accad. Sci. Torino Ser. (2) **67** (1930), 1–15.
[Fi]   S. D. Fisher, *Function theory on planar domains*, Wiley, New York, 1983.
[FS]   C. Fefferman and E. M. Stein, *H^p spaces of several variables*, Acta Math. **129** (1972), 137–193.
[G]   T. Gamelin, *Uniform algebras*, Chelsea, New York, 1984.
[Ga1]   P. Garabedian, *Schwarz's lemma and the Szegö kernel functions*, Trans. Amer. Math. Soc. **67** (1949), 1–35.
[Ga2]   _____, *Distortion of length in conformal mapping*, Duke Math. J. **16** (1949), 439–459.
[Gar]   J. Garnett, *Analytic capacity and measure*, Lecture Notes in Math., vol. 297, Springer-Verlag, Berlin, 1972.
[Gk]   I. E. Garrick, *Potential flow about arbitrary biplane wing sections*, Tech. Rep. No. **542** NACA (1936), 47–75.
[Gr]   H. Grunski, *Lectures on theory of functions in multiply connected domains*, Studia Math.: Skript, vol. 4, Vandenhoeck & Ruprecht, Göttingen, 1978.
[GS]   P. Garabedian and M. Schiffer, *Identities in the theory of conformal mapping*, Trans. Amer. Math. Soc. **65** (1949), 187–238.
[H]   H. Hancock, *Lectures on the theory of elliptic functions*, Dover, New York, 1958.
[Ha]   S. Ya. Havinson, *Analytic capacity of sets, joint nontriviality of various classes of analytic functions and the Schwarz lemma in arbitrary domains*, Mat. Sb. **54** (1961), 3–50. (Russian)
[He1]   D. A. Hejhal, *Theta functions, kernel functions and abelian integrals*, Mem. Amer. Math. Soc., vol. 129, Amer. Math. Soc., Providence, RI, 1972.
[He2]   _____, *Linear extremal problems for analytic functions with interior side conditions*, Ann. Acad. Sci. Fenn. **586** (1974), 5–36.
[HS]   N. S. Hawley and M. Schiffer, *Half-order differentials on Riemann surfaces*, Acta Math. **115** (1966), 199–236.
[J1]   P. W. Jones, *Square functions, Cauchy integrals, analytic capacity and harmonic measure*, Harmonic Analysis and Partial Differential Equations, Lecture Notes in Math., vol. 1384, Springer-Verlag, Berlin, 1989, pp. 24–68.
[J2]   _____, *Rectifiable sets and the traveling salesman problem*, Invent. Math. **102** (1990), 1–16.
[JM]   P. W. Jones and T. Murai, *Positive analytic capacity but zero Buffon needle probability*, Pacific J. Math. **133** (1988), 99–114.
[K]   M. Kaku, *Introduction to superstrings*, Springer-Verlag, New York, 1988.
[Ko]   Y. Komatu, *Die Geschwindigkeitspotentiale und die Kutta-Joukowskischen Bedingungen für die Strömungen in vielfach zusammenhängenden Gebieten*. I, II, Proc. Imperial Acad. Tokyo **21** (1954), 6–15, 83–93.
[Kor]   J. Korevaar, *Polynomial and rational approximation in the complex domain*, Aspect of Contemporary Complex Analysis, Academic Press, London, 1980, pp. 251–292.
[L]   N. S. Landkof, *Foundations of modern potential theory*, Springer-Verlag, Berlin, 1972.
[La]   P. D. Lax, *Reciprocal extremal problems in function theory*, Comm. Pure Appl. Math. **8** (1955), 437–453.
[M]   B. B. Mandelbrot, *The fractal geometry and nature*, Freeman, San Francisco, 1982.
[Ma]   D. E. Marshall, *Removable sets for bounded analytic functions*, Linear and Complex Analysis Problem Book, Lecture Notes in Math., vol. 1043, Springer-Verlag, Berlin, 1984, pp. 485–490.
[Mat1] P. Mattila, *A class of sets with positive length and zero analytic capacity*, Ann. Acad. Sci. Fenn. Ser. AI **10** (1985), 387–395.
[Mat2] _____, *Smooth type, null-sets for integral geometric measure and analytic capacity*, Ann. of Math. **123** (1986), 303–309.

[Me1] Y. Meyer, *Wavelets and operators*, Analysis at Urbana, Vol. I, Lecture Note Ser., vol. 137, London Math. Soc., Cambridge, 1989, pp. 256–365.
[Me2] _____, *Ondelettes et opérateurs*, Hermann, Paris, 1990.
[Mi1] L. M. Milne-Thomson, *Theoretical aerodynamics*, 4th ed., Dover, New York, 1966.
[Mi2] _____, *Theoretical hydrodynamics*, 5th ed., Macmillan, London, 1968.
[Mu1] T. Murai, *A real variable method for the Cauchy transform, and analytic capacity*, Lecture Notes in Math., vol. 1307, Springer-Verlag, Berlin, 1988.
[Mu2] _____, *The power 3/2 appearing in the estimate of analytic capacity*, Pacific J. Math. **143** (1990), 313–340.
[Mu3] _____, *A formula for analytic separation capacity*, Kodai Math. J. **13** (1991), 265–288.
[Mu4] _____, *Analytic capacity for two segments*, Nagoya Math. J. **122** (1991), 19–42.
[Mu5] _____, *The arc-length variation of analytic capacity and a conformal geometry*, Nagoya Math. J. **125** (1992), 151–216.
[Mu6] _____, *Analytic capacity for arcs*, Proc. Internat. Congress Math., (Kyoto, 1990), Amer. Math. Soc., Providence, RI, 1991, pp. 901–911.
[Mu7] _____, *The complex velocity fields induced by several thin obstacles*, J. Math. Pures Appl. **72** (1993), 181–211.
[Mu8] _____, *The role of the Hilbert transform in 2-dimensional aerodynamics*, Approximation by Solutions of Partial Differential Equations, NATO ASI Ser. 365, Kluwer Academic, Dordrecht, 1992, pp. 141–154.
[N] Z. Nehari, *Conformal mappings*, Dover, New York, 1975.
[O] A. G. O'Farrell, *Capacities in function theory*, Potential Theory (Nagoga, 1990), De Gruyter, Berlin, 1992, pp. 93–105.
[P] Ch. Pommerenke, *Über die analytische Kapazität*, Arch. Math. **11** (1960), 270–277.
[R] R. S. Rumely, *Capacity theory on algebraic curves*, Lecture Notes in Math., vol. 1378, Springer-Verlag, Berlin, 1989.
[S] S. Saito, *Theory of reproducing kernels and its applications*, Pitman Res. Notes in Math. Ser., vol. 189, Longman Sci. & Tech., New York, 1988.
[Sa] L. S. Santaló, *Integral geometry and geometric probability*, Addison-Wesley, London, 1976.
[Sas] T. Sasaki, *Applications of conformal mappings*, Fuzanbo, Tokyo, 1939. (Japanese)
[Sc1] M. Schiffer, *On various types of orthogonalization*, Duke Math. J. **17** (1950), 329–366.
[Sc2] _____, *Variational methods in the theory of Riemann surface*, Contributions to the Theory of Riemann Surfaces, Ann. of Math. Stud., no. 30, Princeton Univ. Press, Princeton, NJ, 1953, pp. 15–30.
[Sc3] _____, *Some recent developments in the theory of conformal mappings*, Dirichlet's principle, Pure Appl. Math., vol. 3, Interscience, New York, 1967, pp. 249–318.
[Sch] H. Schlichting, *Boundary-layer theory*, 7th ed., McGraw-Hill, New York, 1987.
[Sm] E. P. Smith, *The Garabedian function of an arbitrary compact set*, Pacific J. Math. **51** (1974), 289–300.
[Su1] N. Suita, *On a metric induced by analytic capacity*, Kodai Math. Sem. Rep. **25** (1973), 215–218.
[Su2] _____, *On a metric induced by analytic capacity*. II, Kodai Math. Sem. Rep. **27** (1976), 159–162.
[Su3] _____, *On subadditivity of analytic capacity for two continua*, Kodai Math. J. **7** (1984), 73–75.
[Su4] _____, *Modern function theory*. II: *Theory of conformal mapping*, Morikita, Tokyo, 1977. (Japanese)
[SH] M. Schiffer and N. S. Hawley, *Connections and conformal mappings*, Acta Math. **107** (1962), 175–274.
[SO] L. Sario and K. Oikawa, *Capacity functions*, Springer-Verlag, Berlin, 1969.
[SS] M. Schiffer and D. C. Spencer, *Functionals of finite Riemann surfaces*, Princeton Univ. Press, Princeton, NJ, 1954.
[T] M. Tsuji, *Potential theory in modern function theory*, Maruzen, Tokyo, 1959.
[V] A. G. Vitushkin, *Example of a set of positive length but zero analytic capacity*, Dokl. Akad. Nauk SSSR **127** (1959), 246–249. (Russian)
[V2] _____, *Analytic capacity of sets in problems of approximation theory*, Uspehi Mat. Nauk **22-26** (1967), 141–199. (Russian)
[Z] L. Zalcman, *Analytic capacity and rational approximation*, Lecture Notes in Math., vol. 50, Springer-Verlag, Berlin, 1986.

Translated by TAKAFUMI MURAI

Laplacians on Self-Similar Sets
—Analysis on Fractals

Jun Kigami

Preliminary remarks

Some people ask "Is there any definition of fractal?". This is the most controversial issue on "fractal" in mathematics. But wait a moment please. Does every term in mathematics have a mathematical definition? How about "algebraic", which is used as in "algebraic geometry", "algebraic topology" or "algebraic analysis". In these examples, its definition would be "to express objects in some appropriate algebra and ...". But this is not a mathematical definition but one of those in an encyclopedia. In fact, it makes no sense to give a mathematical definition to the word "algebraic". Fractal, used in terms such as "fractal geometry" or "analysis on fractal", is the same kind of notion as "algebra" and we do not need a mathematical definition of fractal.

So, what kind of notion is fractal? In *Fractal Geometry of Nature*, Mandelbrot, the pioneer of fractal, gave the following example by Richardson. Let us measure the length of a curve. Dividing a given curve by yardsticks whose length are ε, let $N(\varepsilon)$ be the number of yardsticks we need (cf. Figure 1). Then the approximation of the length of the curve is $L(\varepsilon) = \varepsilon N(\varepsilon)$. Theoretically, the length of the curve is given by $\lim_{\varepsilon \to 0} L(\varepsilon)$. Richardson applied this method to the actual data of the coast lines of Great Britain, Australia, and so on. He plotted the obtained data in X–Y axes where $X = \log_{10} \varepsilon$ and $Y = \log_{10} L(\varepsilon)$ and found that, with considerable accuracy, the data lie on a straight line,

$$Y = (1 - D)X + A,$$

where $A > 0$ and $D = 1.2 \sim 1.3$ depending on coasts. Consequently, $L(\varepsilon) \approx \varepsilon^{1-D}$ as $\varepsilon \to 0$. Hence the length of the coast line should be infinite. From these examples, Mandelbrot insisted that the shape of a coast line is expressed not by an ordinary curve which has finite length but by a curve which satisfies $L(\varepsilon) \approx \varepsilon^{1-D}$.

Traditionally, when we are modelling an object in nature, we always choose smooth figures which have volume, area or length, for example, planes, circles and balls. Even if the shape is complicated, it has been considered to be expressible as a combination of smooth figures. The underlying idea is "However complicated the totality is, it is an aggregation of small simple pieces" or, in other words, "the complexity of the totality $\gg$ the complexity of a part". Mandelbrot pointed out that

---

1991 *Mathematics Subject Classification*. Primary 28A78.
This article originally appeared in Japanese in Sûgaku **44** (1) (1992), 13–28.

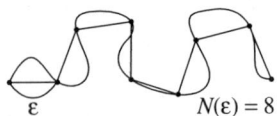

FIGURE 1. How to measure the length of a curve

this is not true. And he introduced the notion of fractal that should be the proper figures by which to express the nature. The typical example is a self-similar set where "the complexity of totality = the complexity of a part".

Mathematicians, however, had been familiar with these kinds of objects like self-similar sets, curves without length or nonintegral dimensions. Early in the 20th century, these objects were found as "pathological" counterexamples to the ordinary smooth calculus and geometry. In fact, almost all of the examples in Mandelbrot's book were found and studied by mathematicians before him. In this sense, the criticism that he has few mathematical results is partially correct. But his true contribution is to point out that those "pathological" objects, which had no positive meaning before, are the proper things by which to describe nature. Philosophically, this is more important than proving theorems. Since then, those classical objects given a new name "fractal" have become one of the most popular subjects in natural sciences.

First, in the study of fractal in mathematics, the main topics were how to generate fractals, for example, self-similar sets, and how to calculate their dimensions, namely, fractal geometry. Supported by classical geometric measure theory, there have been many results in fractal geometry, which we can see in Falconer [**Fa**] and Hata [**Hat2**].

Another direction is complex dynamical systems where beautiful figures of Julia sets and Mandelbrot's set have been attracting mathematicians. This subject originated with French mathematicians Fatou and Julia in the late 19th century. Now quite a profound theory has been developed. Refer to Shishikura [**S**] for recent developments in complex dynamical systems.

By the way, both in fractal geometry and in complex dynamical systems, the main interests are geometrical properties of fractal sets. Is there no other standpoint? The various figures in fractal geometry are models of the objects in nature. On these objects there must occur physical phenomena, for example, waves and diffusions. How can we describe such physical phenomena on fractals? Ordinary analysis can not take care of this problem because it deals only with smooth objects like domains in $\mathbb{R}^n$ and smooth manifolds. We need "analysis on fractals" to describe physical phenomena on fractal nature. To begin with, we should know what "Laplacians" on fractals are in order to describe waves and diffusions. In this paper, we will survey recent developments in this new area "analysis on fractals" from the author's viewpoint.

1. Introduction

The first step in the study of analysis on fractals is to describe diffusions and waves on self-similar sets. In other words, we should formulate "Laplacians" on self-similar sets. In this direction, the pioneering work was the construction of a diffusion process on the Sierpinski gasket by Kusuoka [**Kus1**] and Goldstein [**G**]. Their diffusion process is a scaling limit of random walks on the graphs which approximate the Sierpinski gasket. Refer to Figure 2-a and Definition 2.1 for these graphs. Furthermore, Barlow-

Perkins [**BP**] obtained a detailed estimate of the probability transition density of this diffusion process that is called the "Brownian motion on the Sierpinski gasket". The essential idea of these works is as follows. "In general, it is difficult to consider the notion of derivatives on a fractal. We may, however, construct a diffusion process as a kind of a scaling limit of random walks on graphs which approximate the fractal. The "Laplacian" should be the infinitesimal generator of such a diffusion process." From this direction called the probabilistic approach, Lindstrøm [**Li1**] extended the methods in [**Kus1, G** and **BP**] and obtained "Brownian motions" on a class of highly symmetric self-similar sets named nested fractals. Refer to §7.

On the other hand, Kigami [**Ki1**] defined the "Laplacian" on the Sierpinski gasket directly as a limit of natural difference operators and studied harmonic functions, an expression of solutions of the Poisson equation and the counterpart of Gauss-Green's formula. Later, using these results, Fukushima-Shima [**FS**] and Shima [**Sh1**] determined the eigenvalues and the eigenfunctions for the standard Laplacian on the Sierpinski gasket. This direct approach for constructing Laplacians is called the analytic approach or potential theoretic approach. Much work has been done from this viewpoint, for example, [**Ki3, Me, Kus2 , Ki4**].

These two approaches deal with the same problems from different aspects and the results are complementary. In this paper, we will give a survey on the theory of Laplacians on the Sierpinski gasket mainly from the analytical point of view. Refer to the surveys by Barlow [**B1**] and Lindstrøm [**Li2**] and also the lecture note by Kusuoka [**Kus3**] for the probabilistic approach.

The organization of this paper is as follows. In §2, we will define the Sierpinski gasket and a sequence of graphs $\{V_m\}_{m \geq 0}$ that approximate the Sierpinski gasket and study the discrete Laplacian on those graphs. Laplacians on the Sierpinski gasket will be defined as scaling limits of those discrete Laplacians. In doing so, a renormalization invariance of those discrete Laplacians plays an important role. We will explain this notion of renormalization of difference operators in §3. In §4 and §5, we will define Laplacians, Green's function, and the Dirichlet form on the Sierpinski gasket and state some results on the solution of the Poisson equation and Gauss-Green's formula. In §6, we will introduce the results on the eigenvalue problem of the standard Laplacian on the Sierpinski gasket by [**FS**] and [**Sh1**]. Moreover, in §7, we will extend our theory on the Sierpinski gasket to cover general finitely ramified self-similar sets. Finally, in §8, we will mention, briefly, some recent studies in analysis on fractals.

2. Sierpinski gasket and difference operators

In this section, we will define the Sierpinski gasket (S.G. for short) as a limit of a sequence of finite graphs (V_m, E_m)'s, where the V_m's are the vertices and the E_m's are the edges. On each graph (V_m, E_m), we can find a natural Laplacian as a difference operator. Also later in this section, we will explain how to construct a Laplacian on the S.G. as a limit of the Laplacians on (V_m, E_m).

DEFINITION 2.1. Let p_1, p_2, p_3 be the vertices of an equilateral triangle whose length of each side is 1. Then
(1) For $i = 1, 2, 3$, we define $F_i \colon \mathbb{R}^2 \to \mathbb{R}^2$ by $F_i(x) = \frac{1}{2}(x - p_i) + p_i$. F_i is the contracting similitude which fixes p_i.
(2) For $w = w_1 w_2 \cdots w_m \in \{1, 2, 3\}^m$, we define $F_w \colon \mathbb{R}^2 \to \mathbb{R}^2$ by $F_w = F_{w_1} \circ F_{w_2} \circ \cdots \circ F_{w_m}$.

(3) For $m = 1, 2, \ldots,$

$$V_m = \bigcup_{w \in \{1,2,3\}^m} F_w(\{p_1, p_2, p_3\}),$$

$$E_m = \bigcup_{w \in \{1,2,3\}^m} \{(F_w(p_1), F_w(p_2)), (F_w(p_2), F_w(p_3)), (F_w(p_3), F_w(p_1))\}.$$

In particular, $V_0 = \{p_1, p_2, p_3\}$ and $E_0 = \{(p_1, p_2), (p_2, p_3), (p_3, p_1)\}$.
For $m = 0, 1, 2$, see Figure 2-a.

DEFINITION 2.2. The compact set defined by $K = \overline{\bigcup_{m \geq 0} V_m}$ is called the Sierpinski gasket.

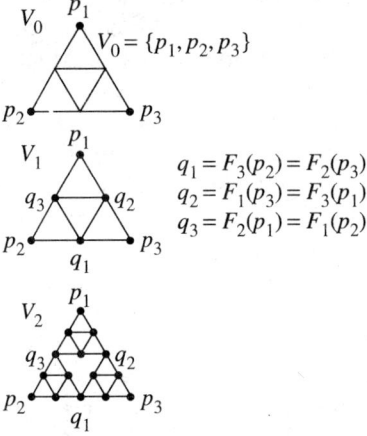

FIGURE 2-a

FIGURE 2-b. The Sierpinski gasket

The Sierpinski gasket K is also characterized as the unique nonempty compact set that satisfies $K = \bigcup_{i=1,2,3} F_i(K)$. The Hausdorff dimension of K is known to be $\frac{\log 3}{\log 2}$. Refer to Hata [**Hat2**] about properties of the Sierpinski gasket as a self-similar set.

Here we prepare some notation.

NOTATION. Let U and V be sets.

(1) $l(V) = \{f | f : V \to \mathbb{R}\}$. We use f_p or $(f)_p$ to denote the value of $f \in l(V)$ at $p \in V$. For $p \in V$, $\chi_p \in l(V)$ is defined by

$$\chi_p(q) = \begin{cases} 1 & \text{for } q = p, \\ 0 & \text{otherwise.} \end{cases}$$

(2) Let $A : l(V) \to l(V)$ be a linear map. Then we use A_{pq} or $(A)_{pq}$ to denote the value $(A\chi_q)_p$ for $q \in V$ and $p \in U$. Note that $\sum_{q \in V} A_{pq} f_q = (Af)_p$.

Now we define the discrete Laplacian on the graph (V_m, E_m) as follows.

DEFINITION 2.3. For $f \in l(V_m)$ and $p \in V_m$,

$$H_{m,p} f = \sum_{q \in V_{m,p}} (f(q) - f(p)),$$

where $V_{m,p} = \{q : q \in V_m, (q,p) \in E_m \text{ or } (p,q) \in E_m\}$.

$V_{m,p}$ is the collection of the neighboring vertices of p in V_m. As is shown in Figure 2-a, we can see that $\#(V_{m,p})$ equals 2 if $p \in V_0$ and 4 otherwise. Thus the points in V_0 are somewhat exceptional. We will think of V_0 as the boundary of the Sierpinski gasket. We will define $V_m^o = V_m \setminus V_0$.

Now, how can we construct a "Laplacian" on the S.G. from the differences $H_{m,p}$'s? For a reference, let us recall the way to describe the Laplacian $\Delta = d^2/dx^2$ on $\mathbb{R}$ as a limit of differences; that is,

(2.1) $$(\Delta f)(x) = \lim_{h \to 0} h^{-2} (f(x+h) + f(x-h) - 2f(x)).$$

As an analogy from the above fact, we give a tentative definition of the Laplacian on the S.G. For given $\alpha > 0$,

DEFINITION 2.4. Let $C(K)$ be the collection of real-valued continuous functions on the S.G. For $f \in C(K)$, if there exists $\varphi \in C(K)$ such that

$$\lim_{m \to \infty} \sup_{p \in V_m^o} |\alpha^m H_{m,p} f - \varphi(p)| = 0$$

then we define $\Delta^{(\alpha)} f = \varphi$. The domain of $\Delta^{(\alpha)}$ is denoted by $\mathscr{D}^{(\alpha)}$.

The first problem is to determine the proper value of α. In the case of the 1-dimensional Laplacian shown in (2.1), we choose the scaling factor h^{-2} with respect to the width of the difference h. This also corresponds to the fact that the Laplacian is the differential operator of order 2. From this point of view, the proper value seems $\alpha = 4$ because the distance of the neighboring vertices in (V_m, E_m) is 2^{-m}. The correct choice is, however, $\alpha = 5$ instead of 4. In fact, we can see that

(1) For $0 < \alpha < 5$, $\text{Ker}\,\Delta^{(\alpha)}$ is dense in $C(K)$.

(2) For $5 < \alpha$, $\mathscr{D}^{(\alpha)}$ is a 3-dimensional subspace of $C(K)$ and for all $f \in \mathscr{D}^{(\alpha)}$, $\Delta^{(\alpha)} f \equiv 0$.

As a matter of fact, the difference operator $H_{m,p}$ is invariant under a kind of renormalization, and the eigenvalue of the renormalization equation, which is a non-linear equation, determines the proper value of α. We will give details about this renormalization in the next section. By virtue of the renormalization invariance, we can obtain the following characterization of harmonic functions.

DEFINITION 2.5. $f \in C(K)$ is called a harmonic function on K if and only if $H_{m,p} f = 0$ for all $m \geq 1$ and all $p \in V_m^o$.

The boundary value problem of harmonic functions is solved as follows.

THEOREM 2.6. *For given $\rho \in l(V_0)$, there exists a unique harmonic function f such that $f|_{V_0} = \rho$. Furthermore, for $w = w_1 w_2 \cdots w_m \in \{1, 2, 3\}^m$,*

$$A_{w_m} \cdots A_{w_2} A_{w_1} \rho = f|_{F_w(V_0)}$$

where

$$A_1 = \frac{1}{5}\begin{pmatrix} 5 & 0 & 0 \\ 1 & 2 & 1 \\ 1 & 1 & 2 \end{pmatrix}, \quad A_2 = \frac{1}{5}\begin{pmatrix} 2 & 1 & 1 \\ 0 & 5 & 0 \\ 1 & 1 & 2 \end{pmatrix}, \quad A_3 = \frac{1}{5}\begin{pmatrix} 2 & 1 & 1 \\ 1 & 2 & 1 \\ 0 & 0 & 5 \end{pmatrix}.$$

The above result implies an algorithm to determine the values of a harmonic function on V_m inductively from $m = 0$. By the way, the equations in Definition 2.5 are formally overdetermined. For example, calculating the values on V_2 from the boundary values on V_0, the number of the unknowns is $\#(V_2^o) = 12$, whereas the equations are $H_{1,p} f = 0$ for $p \in V_1^o$ and $H_{2,p} f = 0$ for $p \in V_2^o$ and hence the total number equals 15. By virtue of the renormalization invariance, however, these 15 equations are not independent and we can obtain Theorem 2.6.

Also, we can see that the harmonic functions form the kernel of the Laplacian.

THEOREM 2.7. *$f \in C(K)$ is a harmonic function if and only if $f \in \mathscr{D}^{(5)}$ and $\Delta^{(5)} f \equiv 0$.*

3. Renormalization of difference operators

As remarked in the previous section, the renormalization invariance plays an important role in constructing Laplacians on the S.G. as scaling limits of the difference operators on V_m's. Furthermore, this notion of "renormalization invariant difference operator" is the key to construct Laplacians on a general class of self-similar sets.

As the domain of renormalization, we define a collection of difference operators on V_0, $\mathscr{H}$ by

$$\mathscr{H} = \left\{ D : D = \begin{pmatrix} -(\alpha + \beta) & \alpha & \beta \\ \alpha & -(\alpha + \gamma) & \gamma \\ \beta & \gamma & -(\beta + \gamma) \end{pmatrix}, \alpha, \beta, \gamma > 0 \right\}.$$

For $D \in \mathscr{H}$, the corresponding difference at $p_1 \in V_0$ is

$$(Df)_{p_1} = -(\alpha + \beta) f(p_1) + \alpha f(p_2) + \beta f(p_3).$$

The differences at p_2 and p_3 are also given similarly.

Now for $D \in \mathcal{H}$, we define a difference operator $H_m(D)$ on V_m as follows.

DEFINITION 3.1. For $w = w_1 w_2 \cdots w_m \in \{1, 2, 3\}^m$, we define $R_w : l(V_m) \to l(V_0)$ by $R_w f = f \circ F_w$. For $D \in \mathcal{H}$, $H_m(D) : l(V_m) \to l(V_m)$ is defined by

$$H_m(D) = \sum_{w \in \{1,2,3\}^m} {}^t R_w D R_w.$$

For example, let

$$D_* = \begin{pmatrix} -2 & 1 & 1 \\ 1 & -2 & 1 \\ 1 & 1 & -2 \end{pmatrix};$$

we have $(H_m(D_*)f)_p = H_{m,p}f$.

Since V_m is the disjoint union of V_0 and V_m^o, $H_m(D)$ can be divided into four parts as

$$H_m(D) = \begin{pmatrix} T_m & {}^t J_m \\ J_m & X_m \end{pmatrix} \begin{pmatrix} f|_{V_0} \\ f|_{V_m^o} \end{pmatrix}$$

where $T_m : l(V_0) \to l(V_0)$, $J_m : l(V_0) \to l(V_m^o)$, and $X_m : l(V_m^o) \to l(V_m^o)$. (In fact, T_m, J_m and X_m depend on D and we should write $T_m(D), J_m(D)$ and $X_m(D)$. For ease of notation, however, we use T_m, J_m and X_m.) Also, we write $T_1 = T$, $J_1 = J$, and $X_1 = X$.

PROPOSITION 3.2. *For $D \in \mathcal{H}$, let*

$$\mathcal{S}_m(D) = T_m - {}^t J_m X_m^{-1} J_m.$$

Then (1) *For $m = 1, 2, 3, \ldots$, $\mathcal{S}_m(D) \in \mathcal{H}$.*
(2) *For $a > 0$, $\mathcal{S}_m(aD) = a\mathcal{S}_m(D)$.*
(3) *For $m, n > 0$, $\mathcal{S}_n \circ \mathcal{S}_m = \mathcal{S}_{n+m}$.*

As a result, $\mathcal{S} = \mathcal{S}_1$ is a homogeneous transformation from $\mathcal{H}$ to itself and $\mathcal{S}_m = \mathcal{S} \circ \mathcal{S} \circ \cdots \circ \mathcal{S}$.

DEFINITION 3.3. $\mathcal{S} : \mathcal{H} \to \mathcal{H}$ is called the renormalization of the difference operator. $D \in \mathcal{H}$ is said to be a renormalization invariant difference operator if and only if there exists $\lambda > 0$ such that $\mathcal{S}(D) = \frac{1}{\lambda} D$.

The following is a known result on the existence, uniqueness and stability of renormalization invariant difference operators.

THEOREM 3.4. *Let $\mathcal{H}_0 = \{D : D \in \mathcal{H}, \text{trace } D = -6\}$ and let $\tilde{\mathcal{S}}(D) = \frac{-6\mathcal{S}(D)}{\text{trace } \mathcal{S}(D)}$. Then*
(1) *$\tilde{\mathcal{S}}$ has the unique fixed point D_*.*
(2) *D_* is stable; that is, for all $D \in \mathcal{H}_0$, $\tilde{\mathcal{S}}^m(D) \to D_*$ as $m \to \infty$, where $\tilde{\mathcal{S}}^m(D) = \tilde{\mathcal{S}} \circ \tilde{\mathcal{S}} \circ \cdots \circ \tilde{\mathcal{S}}$.*

For the proof of the above theorem, refer to [**B2** and **OSS**] for (1) and [**OSS**] for (2).

Essentially the same kinds of renormalizations on self-similar sets were also introduced by [**HHW, Kus2** and **Li1**] in different settings.

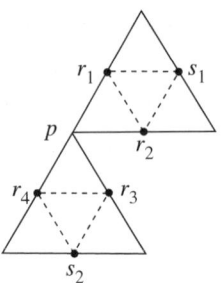

FIGURE 3

Now, how far can we go by using the renormalization invariance of D_*? What we know exactly is

(3.1) $$D_* = \lambda(T - {}^tJX^{-1}J),$$

where $\lambda = \frac{5}{3}$. Indeed, it is not an exaggeration to say that all of the main results in the following sections are essentially the consequence of the renormalization invariance of D_*. For instance, we will show that the notion of harmonic function in the preceding section is well defined.

PROPOSITION 3.5. *For $i = 1, 2, 3$,*

$$3H_{0,p_i}f = 5H_{1,p_i}f + 2\sum_{i \neq j} H_{1,q_j}f + H_{1,q_i}f$$

for all $f \in l(V_1)$, where $q_1 = F_2(p_3) = F_3(p_2)$, $q_2 = F_3(p_1) = F_1(p_3)$ and $q_3 = F_1(p_2) = F_2(p_1)$. (See Figure 2-a.)

PROOF. For $f \in l(V_1)$, let $f_0 = f|_{V_0}$ and let $f_1 = f|_{V_1^o}$. Then by (3.1),

$$D_* f_0 = \lambda(Tf_0 - {}^tJX^{-1}Jf_0)$$
$$= \lambda(Tf_0 + {}^tJf_1) - \lambda {}^tJX^{-1}(Jf_0 + Xf_1).$$

Substituting the following for the above equality, we get the required equations.

$$D_* f_0 = \begin{pmatrix} H_{0,p_1}f \\ H_{0,p_2}f \\ H_{0,p_3}f \end{pmatrix}, \quad Tf_0 + {}^tJf_1 = \begin{pmatrix} H_{1,p_1}f \\ H_{1,p_2}f \\ H_{1,p_3}f \end{pmatrix},$$

$$-{}^tJX^{-1} = \begin{pmatrix} 1 & 2 & 2 \\ 2 & 1 & 2 \\ 2 & 2 & 1 \end{pmatrix}, \quad Jf_0 + Xf_1 = \begin{pmatrix} H_{1,q_1}f \\ H_{1,q_2}f \\ H_{1,q_3}f \end{pmatrix}. \quad \square$$

By Proposition 3.5, for $p \in V_m^o$,

$$3H_{m,p}f = 5H_{m+1,p}f + 2\sum_{i=1}^{4} H_{m+1,r_i}f + \sum_{i=1}^{2} H_{m+1,s_i}f$$

where the locations of $r_1, r_2, r_3, r_4, s_1, s_2 \in V_{m+1}\setminus V_m$ are shown in Figure 3.

Now for $f \in l(V_m)$ that satisfies $H_{m,p}f = 0$ for every $p \in V_m^o$, we can extend f to a function $f \in l(V_{m+1})$ so that $H_{m+1,q}f = 0$ for each $q \in V_{m+1}\setminus V_m$. Then using the above equality, we can see that $H_{m+1,p}f = 0$ for any $p \in V_m^o$. Hence Definition 2.5 is justified. Furthermore for given boundary values $f(p_1), f(p_2), f(p_3)$, the equation $Jf|_{V_0} + Xf|_{V_1^o} = 0$ implies

$$\begin{pmatrix} f(q_1) \\ f(q_2) \\ f(q_3) \end{pmatrix} = \frac{1}{5} \begin{pmatrix} 1 & 2 & 2 \\ 2 & 1 & 2 \\ 2 & 2 & 1 \end{pmatrix} \begin{pmatrix} f(p_1) \\ f(p_2) \\ f(p_3) \end{pmatrix}.$$

In general, for $w \in \{1,2,3\}^m$,

$$\begin{pmatrix} f(F_w(q_1)) \\ f(F_w(q_2)) \\ f(F_w(q_3)) \end{pmatrix} = \frac{1}{5} \begin{pmatrix} 1 & 2 & 2 \\ 2 & 1 & 2 \\ 2 & 2 & 1 \end{pmatrix} \begin{pmatrix} f(F_w(p_1)) \\ f(F_w(p_2)) \\ f(F_w(p_3)) \end{pmatrix}.$$

By the above algorithm, we can determine the values of a harmonic function on V_m from the boundary values on V_0.

4. Laplacians, Dirichlet forms, and Green's functions on V_m

In §4 and §5, we will construct Laplacians, Dirichlet forms, and Green's function on the S.G. from the renormalization invariant difference operator D_*.

NOTATION. (1) $H_m = H_m(D_*)$. Note that $H_{m,p}f = (H_m f)_p$. Also as well as in §3,

$$H_m = \begin{pmatrix} T_m & {}^tJ_m \\ J_m & X_m \end{pmatrix},$$

where $T_m : l(V_0) \to l(V_0)$, $J_m : l(V_0) \to l(V_m^0)$ and $X_m : l(V_m^0) \to l(V_m^0)$.

(2) Let $\lambda = \frac{5}{3}$. Then by (3.1), $D_* = \lambda^m(T_m - {}^tJ_m X_m^{-1} J_m)$.

Now, we choose a Borel probability measure μ on K that satisfies the following.
(1) $\mu(A) = 0$ for every finite set A.
(2) $\mu(O) > 0$ for every nonempty open subset $O \subset K$.

For example, let ν be the normalized $\log 3/\log 2$-dimensional Hausdorff measure on K, which is characterized by $\nu(K_w) = 3^{-m}$ for $w \in \{1,2,3\}^m$, then it satisfies the above conditions.

Next we define Laplacians, Dirichlet forms, and Green's functions on the finite graphs (V_m, E_m)'s.

DEFINITION 4.1. (1) (Laplacian) We define $\Delta_\mu^m : l(V_m) \to l(V_m^o)$ by, for $p \in V_m^o$,

$$(\Delta_\mu^m f)_p = \lambda^m (\mu_{m,p})^{-1} H_{m,p} f,$$

where $\mu_{m,p} = \int_K \psi_p^m d\mu$ and ψ_p^m is the m-harmonic function which satisfies $\psi_p^m|_{V_m} = \chi_p$. $f \in C(K)$ is called an m-harmonic function if and only if $f \circ F_w$ is a harmonic function for all $w \in \{1,2,3\}^m$.

(2) (Dirichlet form) For $u, v \in l(V_m)$, let $\mathscr{E}_m(u,v) = -\lambda^{m\,t}u H_m v$. $\mathscr{E}_m(\cdot,\cdot)$ is a nonnegative symmetric bilinear form on $l(V_m)$.

(3) (Green's function) For $p, q \in V_m$, we define $g_m(p, q)$ by

$$g_m(p,q) = \begin{cases} -\lambda^{-m}(X_m^{-1})_{p,q} & \text{if } p, q \in l(V_m^0), \\ 0 & \text{otherwise.} \end{cases}$$

REMARK 1. In case $\mu = \nu$, for $p \in V_m^o$,

$$(\Delta_\nu^m f)_p = \frac{3}{2} 5^m H_{m,p} f.$$

Thus we can see that Δ_ν^m equals $\Delta^{(5)}$ appearing in Definition 2.4 up to constant multiple.

REMARK 2. $\mathscr{E}_m$ and g_m are independent of μ.

The above definitions imply the following relations.

PROPOSITION 4.2. (1) Let $G_\mu^m : l(V_m^o) \to l(V_m)$ be defined by

$$(G_\mu^m f)_p = \sum_{q \in V_m^0} g_m(p,q) f(q) \mu_{m,q}$$

for $f \in l(V_m^o)$ and $p \in V_m$. Then, $-\Delta_\mu^m \circ G_\mu^m$ is the identity from $l(V_m^o)$ to itself. G_μ^m is the counterpart of the Green's operator with respect to Green's function g_m and μ.

(2) $$\mathscr{E}_m(u,v) = \sum_{p \in V_0} u(p)(-\lambda^m H_{m,p} v) - \sum_{p \in V_m^o} u(p)(\Delta_\mu^m v)_p \mu_{m,p}.$$

(3) For $p \in V_m$, let $g_m^p \in l(V_m)$ be defined by $g_m^p(q) = g_m(p, q)$; then $\mathscr{E}_m(g_m^p, g_m^q) = g_m(p, q)$.

Our final goal is to take the limits of the above objects as $m \to \infty$ and to construct the counterparts of them on the S.G. In such a step, we will make good use of the following facts derived from the renormalization invariance of D_*.

PROPOSITION 4.3. (1) For $u \in l(V_{m+1})$,

$$\mathscr{E}_m(u|_{V_m}, u|_{V_m}) \leq \mathscr{E}_{m+1}(u,u).$$

(2) $g_m(p,q) = g_{m+1}(p,q)$ for $p, q \in V_m$.
(3) There exists $C > 0$ such that for all $m \geq 0$, if $u \in l(V_m)$ and $u|_{V_0} = 0$ then $|u|^2 \leq C\mathscr{E}_m(u,u)$, where $|u| = \sup_{p \in V_m} |u(p)|$.

5. Laplacian, Dirichlet form, and Green's function on the S.G.

In this section, we construct Laplacians, Dirichlet forms, and Green's function on the S.G. as the natural limit of the counterparts on V_m's. First we give the definition of Green's function.

DEFINITION 5.1. Let $\bar{g}_m(x,y) = \sum_{p,q \in V_m} g_m(p,q) \psi_p^m(x) \psi_q^m(y)$. Then by Proposition 4.3(2), $\{\bar{g}_m\}_{m \geq 0}$ is uniformly convergent as $m \to \infty$. We denote this limit by g and call g the Green's function on K. Note that $g : K \times K \to \mathbb{R}$ is continuous.

Next we give the definition of Laplacians.

DEFINITION 5.2. For $f \in C(K)$, if there exists $\varphi \in C(K)$ such that

$$\lim_{m \to \infty} \sup_{p \in V_m^0} |(\Delta_\mu^m f)_p - \varphi(p)| = 0$$

then we define $\Delta_\mu f = \varphi$. The domain of Δ_μ is denoted by $\mathscr{D}_\mu$. In particular, Δ_ν is called the standard Laplacian on the Sierpinski gasket.

REMARK. Recalling the remark after Definition 4.1, we can easily see that $\Delta_\nu = \frac{3}{2}\Delta^{(5)}$.

Now we consider the Dirichlet problem of Poisson's equation for this Laplacian Δ_μ. Using Proposition 4.2(1), we can see that

THEOREM 5.3. *For given $\varphi \in C(K)$ and $\rho \in l(V_0)$, there exists a unique $f \in \mathscr{D}_\mu$ which satisfies*

$$\begin{cases} \Delta_\mu f = \varphi, \\ f|_{V_0} = \rho. \end{cases}$$

Furthermore, f is given by

$$f(x) = \sum_{p \in V_0} \rho(p)\psi_p^0(x) - \int_K g(x,y)\varphi(y)\mu(dy).$$

In the above expression for f, $\sum_{p \in V_0} \rho(p)\psi_p^0$ is a harmonic function whose boundary values equal to ρ. Hence the above formula is an analogue of the ordinary formula for solutions of the Dirichlet problem of Poisson's equation on the bounded domain in $\mathbb{R}^n$, where the solutions are expressed by Green's function and harmonic functions.

Next, we construct a Dirichlet form. By virtue of Proposition 4.3(1), the following is well defined.

DEFINITION 5.4. (1) $\mathscr{F} = \{f : f \in l(V_*), \lim_{m \to \infty} \mathscr{E}_m(f|_{V_m}, f|_{V_m}) < \infty\}$, where $V_* = \bigcup_{m \geq 0} V_m$ and, for $u, v \in \mathscr{F}$,

$$\mathscr{E}(u,v) = \lim_{m \to \infty} \mathscr{E}_m(u|_{V_m}, v|_{V_m}).$$

(2) For $f \in \mathscr{F}$ and $p \in V_0$, we define the Neumann derivative of f at p, $(df)_p$ by $(df)_p = \lim_{m \to 0} -\lambda^m H_{m,p} f$.

REMARK. It can be shown that $\lim_{m \to 0} -\lambda^m H_{m,p} f$ exists for $f \in \mathscr{F}$ and $p \in V_0$.

Now, since V_* is dense in K, we can think of $C(K)$ as a subset of $l(V_*)$ through the natural embedding $f \to f|_{V_*}$. Using Proposition 4.3(3), if $f \in \mathscr{F}$ and $f|_{V_0} = 0$, then $|f|^2 \leq C\mathscr{E}(f,f)$, where $|f| = \sup_{p \in V_*} |f(p)|$. This along with Proposition 4.2 implies

THEOREM 5.5. (1) $\mathscr{D}_\mu \subset \mathscr{F} \subset C(K)$ *and $(\mathscr{E}, \mathscr{F})$ is a local regular Dirichlet form on $L^2(K,\mu)$.*

(2) (*Gauss-Green's formula*) For $u \in \mathcal{F}$ and $v \in \mathcal{D}_\mu$,

$$\mathcal{E}(u,v) = \sum_{p \in V_0} u(p)(dv)_p - \int_K u \Delta_\mu v \, d\mu.$$

(3) For $x \in K$, let $g^x \in C(K)$ be defined by $g^x(y) = g(x,y)$. Then $\mathcal{E}(f, g^x) = f(x) - \sum_{p \in V_0} f(p)\psi_p^0(x)$ for all $f \in \mathcal{F}$. In particular $\mathcal{E}(g^x, g^y) = g(x,y)$.

Refer to Fukushima [**Fu1**] for the definition and fundamental properties of Dirichlet forms. By the above results, we can see that there exists a diffusion process whose infinitesimal generator is Δ_μ. Moreover, it follows that the diffusion process is recurrent. For $\mu = \nu$, this diffusion process is essentially the same as the "Brownian motion" on the S.G. constructed by Kusuoka [**Kus1**], Goldstein [**G**] and Barlow-Perkins [**BP**] as the scaling limit of random walks on V_m.

6. Spectral distributions of Laplacians on the S.G

In this section, we will consider the eigenvalue problems of the Laplacian Δ_μ on the S.G. under the following two boundary conditions.

(i) The Dirichlet boundary condition, for $f \in \mathcal{D}_\mu$,

$$(E_D) \quad \begin{cases} \Delta_\mu f = -kf, \\ f|_{V_0} = 0. \end{cases}$$

(ii) The Neumann boundary condition, for $f \in \mathcal{D}_\mu$,

$$(E_N) \quad \begin{cases} \Delta_\mu f = -kf, \\ (df)_p = 0 \quad \text{for all } p \in V_0. \end{cases}$$

For both (E_D) and (E_N), the eigenvalues are nonnegative, of finite multiplicity, and the only accumulation point is ∞. Hence we can define the ith eigenvalue $k_i^*(\mu)$ of (E_*), where $* = D, N$, taking the multiplicity into account. The eigenvalue counting function $\rho_\mu^*(x)$ is defined by $\rho_\mu^*(x) = \#\{i : k_i^*(\mu) \leq x\}$.

By the way, let Ω be a bounded domain in $\mathbb{R}^n$ and let Δ be the ordinary Laplacian on Ω. Then we know the following famous result by Weyl [**W1, W2**].

THEOREM 6.1 (*Weyl's theorem*). *Let k_i be the ith eigenvalue of the Dirichlet eigenvalue problem of $-\Delta$ on Ω; that is,*

$$\begin{cases} \Delta f = -kf, \\ f|_{\partial\Omega} = 0. \end{cases}$$

Also let $\rho(x) = \#\{i : k_i \leq x\}$, then as $x \to \infty$,

$$\rho(x) = (2\pi)^{-n} \mathcal{B}_n |\Omega|_n x^{n/2} + o(x^{n/2})$$

where $|\cdot|_n$ is the n-dimensional Lebesgue measure and $\mathcal{B}_n = |\{x : |x| \leq 1\}|_n$.

REMARK. Weyl proved the above result under some conditions on the domain Ω. Now, it is known that the above result is true for any bounded domain. Refer to the introduction of [**La1**].

Taking the above theorem into account, we pose

PROBLEM. Is there any kind of analogy of Weyl's theorem in the case of the Laplacian Δ_μ on the S.G.? In particular,

(1) Is the exponent $\alpha = \lim_{x\to\infty} \log \rho_\mu^*(x)/\log x$ related to some dimension of the S.G.?

(2) Does the limit $\lim_{x\to\infty} \rho_\mu^*(x)/x^\alpha$ exist? If it does, what is the meaning of this quantity?

In the case $\mu = \nu$, namely, for the standard Laplacian, the eigenvalues and eigenfunctions of (E_D) and (E_N) were completely determined by Shima [**Sh1**] and Fukushima-Shima [**FS**]. The so-called "decimation method" played the key role in their work. (We will state only the results on (E_D) hereafter. For (E_N), there are similar results.)

LEMMA 6.2 (Decimation method). *Let $\Phi(x) = x(5-x)$. Then we have*
(1) *For $f \in l(V_{m+1}^o)$ and $k \neq 6$, if $X_{m+1}f = -kf$, then $X_m f|_{V_m} = -\Phi(k)f|_{V_m}$.*
(2) *For $f \in l(V_m^o)$ and $k \neq 2, 5, 6$, if $X_m f = -\Phi(k)f$, then there exists a unique extension of f, $\tilde{f} \in l(V_{m+1}^o)$, such that $X_{m+1}\tilde{f} = -k\tilde{f}$.*

Since $\Delta_\nu^m f = \frac{3}{2}5^m X_m f$ for $f \in l(V_m)$ satisfying $f|_{V_0} = 0$, Lemma 6.2 gives a relation between the eigenvalues and eigenfunctions of Δ_ν^m and Δ_ν^{m+1}. Using this relation, we can get detailed information on the eigenvalues and the eigenfunctions of $-\Delta_\nu$. For the eigenvalue counting function,

THEOREM 6.3. *Let $d_S = \frac{\log 9}{\log 5}$;*
$$0 < \liminf_{x\to\infty} \rho_\nu^D(x)x^{-d_S/2} < \limsup_{x\to\infty} \rho_\nu^D(x)x^{-d_S/2} < \infty.$$

Comparing with Weyl's theorem, d_S could be a kind of "dimension" of the S.G. In fact, it is called the spectral dimension. The well-known Hausdorff dimension of the S.G. is, however, $d_H = \log 3/\log 2$. Thus, for the S.G., the dimension d_H from the geometrical viewpoint differs from the dimension d_s from the analytical viewpoint. Some partial results were given by Kigami [**Ki4**] to fill in the gap between these two dimensions.

Making use of Lemma 6.2, we can also find a remarkable fact on an eigenfunction.

PROPOSITION 6.4. *There exists a nontrivial eigenfunction φ such that $\operatorname{supp} \varphi = \overline{\{x : \varphi(x) \neq 0\}} \subset K_1 = F_1(K)$.*

Let $u(x, t) = e^{-kt}\varphi(x)$, where $\Delta_\nu \varphi = -k\varphi$, then u is a solution of the diffusion equation $\frac{\partial u}{\partial t} = \Delta_\nu u$. The support of this solution $\overline{\{x : u(x, t) \neq 0\}}$ is contained in K_1 for all $t \geq 0$. Hence the heat never diffuses outside K_1. In the same way, we can construct a solution of the wave equation $\frac{\partial^2 u}{\partial t^2} = \Delta_\nu u$ from φ. In this case, the wave is localized in K_1. This kind of phenomenon is not observed in solutions of the ordinary diffusion or wave equation on $\mathbb{R}^n$.

Thus, by virtue of the decimation method, we can obtain many concrete and interesting results on the eigenvalues and eigenfunctions of Δ_ν. Unfortunately, the decimation method does not work for general μ. Recently, motivated by the method of the Dirichlet-Neumann bracketing on the self-similar sets introduced by Fukushima [**Fu2**], Kigami-Lapidus [**KL**] studied the case where μ is a Bernoulli measure. μ is called a Bernoulli measure if $\mu(K_w) = \mu_{w_1}\mu_{w_2}\cdots\mu_{w_m}$ for $w = w_1 w_2 \cdots w_m \in \{1, 2, 3\}^m$, where $\mu_i = \mu(K_i)$ for $i = 1, 2, 3$.

THEOREM 6.5. *Let μ be a Bernoulli measure. Then for $* = D, N$,*

$$0 < \liminf_{x \to \infty} \rho_\mu^*(x)/x^{d_S/2} \leq \limsup_{x \to \infty} \rho_\mu^*(x)/x^{d_S/2} < \infty$$

where d_S is the unique constant that satisfies

$$(\tfrac{3}{5}\mu_1)^{d_S/2} + (\tfrac{3}{5}\mu_2)^{d_S/2} + (\tfrac{3}{5}\mu_3)^{d_S/2} = 1.$$

Moreover, let $\beta_i = \log \tfrac{3}{5}\mu_i$. Then if the additive group $\sum_{i=1}^{3} \mathbb{Z}\beta_i$ is dense in $\mathbb{R}$, the limit in the above equation exists and

$$\lim_{x \to \infty} \rho_\mu^D(x)/x^{d_S/2} = \lim_{x \to \infty} \rho_\mu^N(x)/x^{d_S/2}.$$

Actually, Fukushima [**Fu2**] studied the spectral distribution of the Laplacians on nested fractals introduced by Lindstrøm [**Li1**]. And Kigami-Lapidus [**KL**] dealt with the Laplacians on p.c.f. self-similar sets introduced by Kigami [**Ki3**].

7. Generalization

In the previous sections, we reviewed the results on the Dirichlet form, Laplacians and Green's function on the Sierpinski gasket. We can also construct these analytical structures on a more general class of self-similar sets. In this section, we will briefly introduce the framework by Kigami [**Ki3**], where Dirichlet forms and Laplacians are constructed on post critically finite (p.c.f. for short) self-similar sets. Post critically finite self-similar sets are a purely topological formulation of "finitely ramified" self-similar sets. In this framework, the key idea is the concept of renormalization invariant difference operators.

DEFINITION 7.1. Let K be a compact metrizable topological space and let S be a finite set.[1] Also, for $i \in S$, let F_i be a continuous injection from K to itself. Then $(K, S, \{F_i\}_{i \in S})$ is called a self-similar structure if there exists a continuous surjection $\pi : \Sigma \to K$ such that $F_i \circ \pi = \pi \circ i$ for every $i \in S$, where $\Sigma = S^{\mathbb{N}}$ is the one-sided shift space and $i : \Sigma \to \Sigma$ is defined by $i(w_1 w_2 w_3 \cdots) = i w_1 w_2 w_3 \cdots$ for each $w_1 w_2 w_3 \cdots \in \Sigma$.

For each $i \in S$, $i : \Sigma \to \Sigma$ is one of the branches of the inverse of the shift $\sigma : \Sigma \to \Sigma$, where $\sigma(\omega_1 \omega_2 \omega_3 \cdots) = \omega_2 \omega_3 \omega_4 \cdots$. Hence by the above definition, if $C_K = \bigcup_{i \neq j}(K_i \cap K_j)$ is empty, then π is bijective and σ can be induced on K. Thus we can think of C_K as the obstruction for inducing σ on K.

DEFINITION 7.2. Let $(K, S, \{F_i\}_{i \in S})$ be a self-similar structure. We define the critical set $\mathscr{C} \subset \Sigma$ and the post critical set $\mathscr{P} \subset \Sigma$ by

$$\mathscr{C} = \pi^{-1}\left(\bigcup_{i \neq j}(K_i \cap K_j)\right) \quad \text{and} \quad \mathscr{P} = \bigcup_{n \geq 1} \sigma^n(\mathscr{C})$$

A self-similar structure is called post critically finite (p.c.f. for short) if and only if $\#(\mathscr{P})$ is finite.

For a p.c.f self-similar structure $(K, S, \{F_i\}_{i \in S})$, we can define V_m in the same way as in Definition 2.1.

---

[1] In this paper, $S = \{1, 2, \ldots, N\}$.

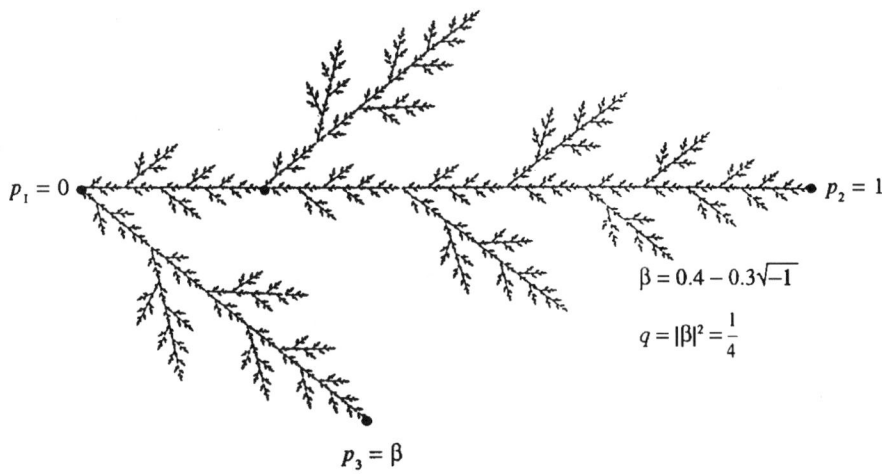

FIGURE 4. The Hata's tree-like set

DEFINITION 7.3. Let $(K, S, \{F_i\}_{i \in S})$ be a p.c.f. self-similar structure. We define V_m for $m \geq 0$ by $V_0 = \pi(\mathcal{P})$ and $V_m = \bigcup_{w \in S^m} F_w(V_0)$ for $m \geq 1$, where $F_w = F_{w_1} \circ F_{w_2} \circ \cdots \circ F_{w_m}$ for $w = w_1 w_2 \cdots w_m \in S^m$.

EXAMPLE 7.4. (1) Let K be the Sierpinski gasket and let F_i be the contractions defined in Definition 2.1. Then $C_K = \{q_1, q_2, q_3\}$ and

$$\mathscr{C} = \{1\dot{2}, 2\dot{1}, 2\dot{3}, 3\dot{2}, 3\dot{1}, 1\dot{2}\} \quad \text{and} \quad \mathscr{P} = \{\dot{1}, \dot{2}, \dot{3}\},$$

where $\dot{k} = kkkk \cdots$. Also we can see $p_k = \pi(\dot{k})$. Therefore $V_0 = \{p_1, p_2, p_3\}$.

(2) We define $F_i : \mathbb{C} \to \mathbb{C}$ for $i \in \{1, 2\} = S$ by $F_1(z) = \beta \bar{z}$ and $F_2(z) = (1 - |\beta|^2)\bar{z} + |\beta|^2$, where $|\beta| < 1$, $|1 - \beta| < 1$ and, $\text{Im }\beta \neq 0$. The unique nonempty compact set K satisfying $K = F_1(K) \cup F_2(K)$ is called the Hata's tree-like set, which was found by Hata [**Hat1**]. In Figure 4, $\beta = \frac{4+3\sqrt{-1}}{10}$. Then $(K, S, \{F_i\}_{i \in S})$ is a p.c.f self-similar structure. In fact, $C_K = \{q\}$, $\mathscr{C} = \{11\dot{2}, 2\dot{1}\}$, and $\mathscr{P} = \{\dot{1}, \dot{2}, 1\dot{2}\}$. Moreover, $V_0 = \{p_1, p_2, p_3\}$ and $V_1 = \{p_1, p_2, p_3, q\}$.

We can define the renormalization of difference operators on p.c.f. self-similar structures in entirely the same way as in the case of the S.G. except that the $F_i(K)$'s may have different sizes in general. For example, $F_1(K)$ and $F_2(K)$ are similar but not congruent in the case of the Hata's tree-like set. Thus for general p.c.f. self-similar structures, we need to introduce a weight factor $r = (r_i)_{i \in S}$.

DEFINITION 7.5. For a difference operator $D : l V_0 \to l(V_0)$ on V_0, we define $H_1 : l(V_1) \to l(V_1)$ by $H_1 = \sum_{i \in S} r_i^{-1} \, {}^t R_i D R_i$, where $R_i f = f \circ F_i$. Furthermore, let

$$\mathscr{S}_r(D) = T - {}^t J X^{-1} J,$$

where

$$H_1 = \begin{pmatrix} T & {}^t J \\ J & X \end{pmatrix}.$$

$\mathscr{S}_r$ is the renormalization of the difference operators. If there exists $\lambda > 0$ such that $\mathscr{S}_r = \frac{1}{\lambda} D$, then D is called a renormalization invariant difference operator. If we get a pair of a weight factor r and a $\mathcal{S}_r$-invariant difference operator D, we can construct a Dirichlet form, Laplacians and Green's function in exactly the same way as in §2–§6 and show the results corresponding to Theorem 5.5. The spectral distribution of such Laplacians on p.c.f. self-similar sets are studied by [**KL**].

EXAMPLE 7.6. For the Hata's tree-like set, let $r = (\alpha, 1 - \alpha^2)$ for $0 < \alpha < 1$, then

$$D = \begin{pmatrix} -(1+\alpha^{-1}) & 1 & \alpha^{-1} \\ 1 & -1 & 0 \\ \alpha^{-1} & 0 & -\alpha^{-1} \end{pmatrix}, \quad \lambda = 1.$$

is the $\mathcal{S}_r$-invariant difference operator.

Now it is important whether there exists a renormalization invariant difference operator for a given p.c.f. self-similar structures or not. Specifically, for a given p.c.f. self-similar structure and a given weight factor r,
  (1) Is there any S_r-invariant difference operator? If there is,
  (2) Is it unique up to constant multiple?
  (3) Is it stable under perturbations?
By Theorem 3.4, the answers are yes for (1), (2), and (3) in the case of the S.G. where $r = (1, 1, 1)$. Lindstrøm [**Li1**] introduced the notion of nested fractals, which are p.c.f. self-similar sets with strong symmetry, and showed the existence of decimation invariant random walks. In our framework, this result says that for a nested fractal, let $r = (1, 1, \ldots, 1)$, then there exists an $\mathcal{S}_r$-invariant difference operator. So far, this is the best result for the above problems.

8. Related works

In this section, we will briefly mention some works concerning analysis on fractals.
  (1) Self-avoiding process on the Sierpinski gasket. Hattori et al. [**HHK, HH, HK**] constructed and studied a process on the S.G. where a path will not intersect itself.
  (2) Diffusions processes on randomized self-similar sets. Hambly [**Ham1** and **Ham2**] constructed and studied diffusion processes and associated Dirichlet forms on randomized Sierpinski gaskets, which are not self-similar sets. Figure 5 shows the first several steps of the construction of a randomized S.G.
  (3) Random Schrödinger operators on nested fractals. Paluba [**P**] studied the integrated densities of states and their Lifschitz type singularities for random Schrödinger operators, which are the Laplacian plus random potentials, on the Sierpinski gasket. Shima [**Sh2**] extended her results to random Schrödinger operators on nested fractals.
  The objects of the above works in (1), (2) and (3) are finitely ramified fractals. Other than these, Kumagai [**Kum**] constructed nonsymmetric diffusion processes on the Sierpinski gaskets. Also, Fujita [**Fj1, Fj2**] studied 1-dimensional generalized diffusion processes whose speed measures are self-similar measures.

FIGURE 5

FIGURE 6

(4) Infinitely ramified fractals. One of the famous examples of infinitely ramified self-similar set is the Sierpinski carpet. Cf. Figure 6 for the first two stages of generating the Sierpinski carpet. Compared to finitely ramified cases, it is more difficult to construct analytical structures such as diffusion processes, Dirichlet forms and Laplacians on infinitely ramified self-similar sets. If we directly extend the method used for the finitely ramified case, the renormalization becomes a nonlinear transformation on infinite dimensional space. Despite such a difficulty, Barlow-Bass [**BB1**] constructed a diffusion process, called the Brownian motion, on the Sierpinski carpet by taking a limit of the Brownian motions on the domains in $\mathbb{R}^2$ approximating the Sierpinski carpet. Furthermore they obtained the existence of local time and the same type estimate of the transition probability density in the case of Sierpinski gasket [**BB2, 3, 4; BBS**]. Also Osada [**O**] obtained an estimate of the spectral dimension of the Sierpinski gasket.

(5) Laplacians on the domain whose boundaries are fractal Let $\Omega \in \mathbb{R}^n$ be a bounded domain. Lapidus et al. [**LF, La2, LM, LP1, LP2**] studied the eigenvalue counting function of the Laplacian on Ω when $\partial\Omega$ is a fractal. The behavior of the leading term determined by Weyl's theorem (Theorem 6.1). According to their result, the asymptotic behavior of the second term is characterized by the dimension of $\partial\Omega$. Refer to the survey by Lapidus [**La1**] for the results in this area.

References

[B1]   M. T. Barlow, *Random walks and diffusion on fractals*, Proc. Internat. Congr. Math. (Kyoto 1990), Math. Soc. of Japan, and Springer, 1991, pp. 1025–1035.

[B2] _____, *Random walks, electrical resistance, and nested fractals*, Asymptotic Problems in Probability Theory: Stochastic Models and Diffusions on Fractals (K. D. Elworthy and N. Ikeda, eds.), Pitman Res. Notes in Math., no. 283, Longman, 1993, pp. 131–157.

[BB1] M. T. Barlow and R. F. Bass, *The construction of Brownian motion on the Sierpinski carpet*, Ann. Inst. H. Poincaré **25** (1989), 225–257.

[BB2] _____, *Local time for Brownian motion on the Sierpinski carpet*, Probab. Theory Related Fields **85** (1990), 91–104.

[BBS] M. T. Barlow, R. F. Bass and J. D. Sherwood, *Resistance and spectral dimension of Sierpinski carpets*, J. Phys. A **23** (1990), L253–L258.

[BB3] M. T. Barlow and R. F. Bass, *On the resistance of the Sierpinski carpet*, Proc. Roy. Soc. London Ser. A**431** (1990), 354–360.

[BB4] _____, *Transition densities for Brownian motion on the Sierpinski carpet*, Probab. Theory Related Fields **91** (1992), 307–330.

[BP] M. T. Barlow and E. A. Perkins, *Brownian motion on the Sierpinski gasket*, Probab. Theory Related Fields **79** (1988), 542–624.

[Fa] K. J. Falconer, *The geometry of fractal sets*, Cambridge Univ. Press, 1985.

[Fj1] T. Fujita, *A fractional dimension, self-similarity and a generalized diffusion operator*, Probabilistic Methods on Mathematical Physics, Proc. of Taniguchi Internat. Sympos. (Katata and Kyoto, 1985), (K. Ito and N. Ikeda eds), Kinokuniya, Tokyo, 1987, pp. 83–90.

[Fj2] _____, *Some asymptotics estimates of transition probability densities for generalized diffusion processes with self-similar speed measures*, preprint.

[Fu1] M. Fukushima, *Dirichlet forms and Markov processes*, North-Holland/Kodansya, 1980.

[FS] M. Fukushima and T. Shima, *On a spectral analysis for the Sierpinski gasket*, Potential Analysis **1** (1992), 1–35.

[Fu2] M. Fukushima, *Dirichlet forms, diffusion processes and spectral dimensions for nested fractals*, Ideas and Methods in Mathematical Analysis, Stochastics and Application, In memory of Hoegh-Krøhn, vol. 1, (Albeverio, Fenstad, Holden, Lindstrøm, eds.), Cambridge Univ. Press, 1992.

[G] S. Goldstein, *Random walks and diffusions on fractals*, Kesten, H. (Ed.) Percolation Theory and Ergodic Theory of Infinite Particle Systems, IMA Math. Appl., vol. 8, Springer, New York, 1987, pp. 121–129.

[Ham1] B. Hambly, *Diffusion on a class of random fractals*, Ph. D. thesis, Univ. of Cambridge, 1990.

[Ham2] _____, *Brownian motion on a homogeneous random fractal*, Probab. Theory Related Fields **94** (1992), 1–38.

[Hat1] M. Hata, *On the structure of self-similar sets*, Japan J. Appl. Math. **3** (1985), 381–414.

[Hat2] _____, *Fractal—on self-similar sets*, Sûgaku **42** (1990), 304–317. (Japanese)

[HHW] K. Hattori, T. Hattori and H. Watanabe, *Gaussian field theories and the spectral dimensions*, Progr. Theoret. Phys. Suppl. **92** (1987), 108–143.

[HHK] K. Hattori, T. Hattori and S. Kusuoka, *Self-avoiding paths on the pre-Sierpinski gasket*, Probab. Theory Related Fields **84** (1990), 1–26.

[HH] K. Hattori and T. Hattori, *Self-avoiding process on the Sierpinski gasket*, Probab. Theory Related Fields **88** (1991), 405–428.

[HK] T. Hattori and S. Kusuoka, *The exponent for mean square displacement of self-avoiding randomwalk on Sierpinski gasket*, Probab. Theory Related Fields **93** (1992), 273–284.

[Hu] J. E. Hutchinson, *Fractals and self-similarity*, Indiana Math. J. **30** (1981), 713–743.

[Ki1] J. Kigami, *A harmonic calculus on the Sierpinski spaces*, Japan J. Appl. Math. **6** (1989), 259–290.

[Ki2] _____, *A calculus on some self-similar sets*, Fractals in the Fundamental and Applied Sciences, (H.-O. Peitgen, J.M. Henriques and L.F. Pendo, eds.), Elsevier Science Publishers B. V., North-Holland, 1991, pp. 239–254.

[Ki3] _____, *Harmonic calculus on p.c.f. self-similar sets*, Trans. Amer. Math. Soc. **335** (1993), 721–755.

[Ki4] _____, *Harmonic metric and Dirichlet form on the Sierpinski gasket*, Asymptotic Problems in Probability Theory: Stochastic Models and Diffusions on Fractals, (K. D. Elworthy and N. Ikeda, eds.), Pitman Res. Notes in Math., no. 283, Longman 1993, pp. 201–218.

[KL] J. Kigami and M. L. Lapidus, *Weyl's problem for the spectral distributions of Laplacians on p.c.f. self-similar fractals*, Comm. Math. Phys. **158** (1993), 93–125.

[Kum] T. Kumagai, *Construction and some properties of a class of non-symmetric diffusion process on the Sierpinski gasket*, Asymptotic Problems in Probability Theory: Stochastic Models and Diffusions on Fractals, (K. D. Elworthy and N. Ikeda, eds.), Pitman Res. Notes in Math., no. 283, Longman, 1993, pp. 219–247.

[Kus1] S. Kusuoka, *A diffusion process on a fractal*, Probabilistic Methods on Mathematical Physics, Proc. Taniguchi Internat. Sympos. (Katata and Kyoto, 1985), (K. Ito and N. Ikeda, eds.), Kinokuniya, Tokyo, 1987, pp. 251–274.

[Kus2] _____, *Dirichlet forms on fractals and products of random matrices*, Publ. Res. Inst. Math. Sci. **25** (1989), 659–680.

[Kus3] _____, *Lecture on diffusion processes on nested fractals*, preprint.

[LF] M. L. Lapidus and J. Fleckinger-Pellé, *Tambour fractal: vers une résolution de la conjecture de Weyl-Berry pour les valeurs propres de laplacien*, C. R. Acad. Sci. Paris Sér. I Math. **306** (1988), 171–175.

[La1] M. L. Lapidus, *Fractal drum, inverse spectral problems for elliptic operators and a partial resolution of the Weyl-Berry conjecture*, Trans. Amer. Math. Soc. **325** (1991), 465–529.

[La2] _____, *Spectral and fractal geometry: from the Weyl-Berry conjecture for the vibrations of fractal drums to the Riemann zeta-function*, Proc. UAB Internat. Conf. on Math. Phys. and Differential Equations (Birmingham, 1990), (C. Bennewitz, ed.), Academic Press, New York, 1992, pp. 151–182.

[LM] M. L. Lapidus and H. Maier, *Hypothése de Riemann, cordes fractales vibrantes et conjecture de Weyl-Berry modifieé*, C. R. Acad. Sci. Paris Sér. I Math. **313** (1991), 19–24.

[LP1] M. L. Lapidus and C. Pomerance, *Fonction zéta de Riemann et conjecture de Weyl-Berry pour les tambour fractals*, C. R. Acad. Sci. Paris Sér. I Math. **310** (1990), 343–348.

[LP2] _____, *The Riemann zeta-function and the one-dimensional Weyl-Berry conjecture for fractal drums*, Proc. London Math. Soc. **66** (1993), 41–69.

[Li1] T. Lindstrøm, *Brownian motion on nested fractals*, Mem. Amer. Math. Soc., no. 420, 1990.

[Li2] _____, *Nonstandard analysis, iterated function systems and Brownian motion on fractals*, preprint, Univ. of Oslo, 1990.

[Li3] _____, *Brownian motion penetrating the Sierpinski gasket*, Asymptotic Problems in Probability Theory: Stochastic Models and Diffusions on Fractals (K. D. Elworthy and N. Ikeda, eds.), Pitman Res. Notes in Math., no. 283, Longman, 1993, pp. 248–278.

[Me] V. Metz, *Potentialtheorie auf dem Sierpinski gasket*, Math. Ann. **289** (1991), 207–237.

[Mo] P. A. P. Moran, *Additive functions of interval and Hausdorff measure*, Proc. Cambridge. Philos. Soc. **42** (1946), 15–23.

[OSS] M. Okada, T. Sekiguchi and Y. Shiota, *Heat kernels on infinite graph networks and deformed Sierpinski gaskets*, Japan J. Appl. Math. **7** (1990), 527–543.

[O] H. Osada, *Isoperimetric dimenstion and etimates on the heat kernels of pre-Sierpinski carpets*, Probab. Theory Related Fields **86** (1990), 469–490.

[P] K. Piertruska-Paluba, *The Lifschitz singularity for the density of states on the Sierpinski gasket*, Probab. Theory Related Fields **89** (1991), 1–33.

[S] M. Shishikura, *Complex dynamical systems on the Riemann sphere*, Sûgaku **41** (1989), pp. 34–48. (Japanese)

[Sh1] T. Shima, *On eigenvalue problems for the random walks on the Sierpinski pre-gaskets*, Japan J. Indust. Appl. Math. **8** (1991), 124–141.

[Sh2] T. Shima, *Lifshitz tails for random Schrödinger operators on nested fractals*, Osaka J. Math. **29** (1992), 749–770.

[Si] W. Sierpinski, *Sur une courbe dont tout point est une point remification*, C. R. Acad. Sci. Paris **160** (1915), 302–305.

[W1] H. Weyl, *Das asymptotisce Verteilungsgesetz der Eigenwerte linearer partieller Differentialgleichungen*, Math. Ann. **71** (1912), 441–479.

[W2] _____, *Über die Abhängigkeit der Eigenschwingungen einer Membran von deren Bergrenzung*, J. Reine Angew. Math. **141** (1912), 1–11.

[YK] M. Yamaguti and J. Kigami, *Some remarks on Dirichlet problem of Poisson equation*, Analyse Mathématique et Application, Gauthier-Villars, Paris, 1988, pp. 465–471.

GRADUATE SCHOOL OF HUMAN AND ENVIRONMENTAL STUDIES, KYOTO UNIVERSITY, KYOTO 606-01, JAPAN

Translated by JUN KIGAMI

Statistical Analysis of Mapped Point Patterns—Present Condition of Theory and Application

Shigeru Mase, Yosihiko Ogata, and Masami Tanemura

1. Introduction

Recently, there have been rapidly growing accumulations of spatial (geometrical) data observed from spatially spread objects. They arise, for example, in environmental sciences, ecology, medical science, seismology, geography, and so on. In order to obtain new knowledge and to make useful forecasts from these data, we must analyze geometrical aspects of the objects involved.

Of the various spatial data, the data of mapped point patterns are the most basic. Imagine positions of animal's nests in their homeland or positions of trees in a forest as examples. They show various typical spatial random patterns. How can these patterns be characterized statistically, and how should we fit particular statistical models to them?

This problem has its origin in ecology and geography, and Japanese researchers have been making important contributions. At present it is one of the central problems of "spatial statistics", a newly coined name of the branch of statistics that deals with geometrical data. We refer readers to the books of Diggle [5], Ripley [41], and Stoyan, et al. [44] for an introduction to this problem.

In this paper we discuss the problem of estimating the interaction force from mapped point pattern data assuming that the point pattern was generated under the influence of an interaction force acting among points. For this purpose we need probability distributions with likelihoods which represent interaction explicitly from both theoretical and practical viewpoints. On this account it is quite natural to conceive the idea of using Gibbsian distributions which have been long used in statistical physics as a model of equilibrium patterns of physical particles. The authors have been studying the statistical estimation problem based on Gibbsian distributions. Nevertheless, the existing results are partial at best, and we can show no consistent theory. Therefore, our aim in this paper is to report the present condition of the theory and invite readers to this interesting and promising subject.

2. Definition of Gibbsian distribution

The concept of a Gibbsian distribution is far from well known in the statistical world, and it is appropriate to state its definition in some detail. Let us give the

---

1991 *Mathematics Subject Classification.* Primary 62F99.
This article originally appeared in Japanese in Sûgaku **44** (3) (1992), 193–204.

definition of continuous state space Gibbsian distributions at first. Let $\mathbf{B}$ (resp., $\mathbf{B}_0$) be the Borel (resp., bounded Borel) sets of a locally compact Hausdorff space S. The set of configurations, that is, locally finite subsets, is denoted by $\mathscr{C}_S$. The number function $N_\Lambda(x) = \#(x \cap \Lambda)$ is defined for each $\Lambda \in \mathbf{B}_0$. The space $\mathscr{C}_S$ is endowed with the smallest σ-algebra $\mathbf{C}_S$ which makes all number functions measurable. In particular, if $S = \mathbf{R}^2$, our main case, we simply write $\mathscr{C}$ and $\mathbf{C}$. For each configuration x, x_A stands for $x \cap A$. Any upper semicontinuous function from $(0, \infty)$ into $(-\infty, +\infty]$ is called a potential function. A potential function Φ is said to be stable if there is a nonnegative constant B such that the inequality

$$V(x) \equiv \frac{1}{2} \sum \{\Phi(|s-t|); s, t \in x, s \neq t\} \geq -B\#x$$

holds for every configuration x. For two configurations x, y, define the local energy $V(x|y)$ of x given the outer configuration y by the following

$$V(x|y) = \# x \cdot z + V(x) + \sum \{\Phi(|s-t|); s \in x, t \in y\},$$

where $z \in \mathbf{R}$ is an arbitrary given number called the chemical potential. We will use the symbols $(x)_n = \{x_1, \ldots, x_n\}$, $d(x)_n = dx_1 \cdots dx_n$, and $r_{ij} = |x_i - x_j|$. The Poisson point process (distribution) ν_Λ with unit intensity defined on the domain $\Lambda \in \mathbf{B}_0$ is the probability measure on $\mathscr{C}_\Lambda$ defined as follows

$$\int f(x) d\nu_\Lambda(x) = e^{-|\Lambda|} \left[f(\varnothing) + \sum_{n=1}^{\infty} \frac{1}{n!} \int_{\Lambda^n} f((x)_n) d(x)_n \right].$$

DEFINITION 1. The grand canonical (local) Gibbsian distribution $\mu_\Lambda(\cdot|y)$ defined on a domain $\Lambda \in \mathbf{B}_0$ and having an outer configuration $y \in \mathscr{C}$ is the probability distribution which has the density

(1) $$\exp(-V(x|y_{\Lambda^c}))/\Xi_\Lambda(y)$$

with respect to the Poisson distribution ν_Λ. The normalizing constant Ξ_Λ called the grand canonical partition function is defined as

$$\Xi_\Lambda(y) \equiv \int_{\mathscr{C}_\Lambda} \exp(-V(x|y_{\Lambda^c})) d\nu_\Lambda(x).$$

The Gibbsian distribution $\mu_\Lambda(\cdot|y)$ actually depends only on y_{Λ^c}. If we consider the conditional distribution given the condition on the number of points n, then we have the following definition.

DEFINITION 2. Let $\mathscr{C}_{n,\Lambda}$ be the set of n-points configurations on Λ. The (n-points) canonical Gibbsian distribution $\mu_{n,\Lambda}$ on a domain Λ is the probability distribution defined on $\mathscr{C}_{n,\Lambda}$ which has the following density with respect to $d(x)_n$:

(2) $$\exp(-V((x)_n))/\Xi_{n,\Lambda}$$

(we do not consider outer configurations). The normalizing constant $\Xi_{n,\Lambda}$ is called the canonical partition function and is defined as

$$\Xi_{n,\Lambda} \equiv \int_{\mathscr{C}_{n,\Lambda}} \exp\{-V((x)_n)\} d(x)_n.$$

DEFINITION 3. A probability distribution defined on $\mathscr{C}$ is called a (global) Gibbsian distribution if its conditional distribution $\mu(\cdot|\mathscr{C}_{\Lambda^c})$ for each $\Lambda \in \mathbf{B}_0$ is given by

$\mu_\Lambda(\cdot|y)$. The distribution is called stationary if it is unchanged under every shift transformation of configurations $\{x_n\} \mapsto \{x_n + t\}$. It is called isotropic if it is unchanged under every rotation of configurations.

The local energy means the total energy associated with configurations. As equation (1) shows, configurations with larger (resp., smaller) local energy are more (less) unstable and have smaller (resp., larger) existing probabilities. The chemical potential can be interpreted as the parameter which governs the degree of in and out (or birth and death) of points from a domain. If z is large, then configurations with many points have larger energies and become more unstable. An attractive feature of Gibbsian distributions from the point of view of a statistical model is that we can generate various fairly complicated and realistic point patterns starting from simple potential functions and that potential functions can admit a clear interpretation as the degree of interaction of forces acting among points. On the other hand, Gibbsian distributions are essentially complicated distributions having complex dependencies, and its theoretical analysis is very difficult. This complexity persists also in the applications as the difficulty of numerical calculation of partition functions. We refer readers to Preston [38] for the basic existence and uniqueness results of Gibbsian distributions. The phenomenon called phase transitions (that there can be multiple global Gibbsian distributions for some potential functions) is one more hindrance for statistical analysis.

We can also define Gibbsian distributions on the lattice $\mathbf{Z}^2$ with values in a general space. Although, we refer the readers to Preston [38] and Georgii [10] for details, we remark that we can associate each continuous state space Gibbsian distribution with its discretized lattice Gibbsian distribution having a general state space. This enables us to study properties of continuous state space distributions from those of lattice distributions, and this is the basis of the discussion in §4.

Divide $\mathbf{R}^2$ into square grids Λ_i with center $i \in \mathbf{Z}^2$. Let ν_i be the Poisson distribution with unit intensity on $(\mathscr{E}_{\Lambda_i}, \mathbf{C}_{\Lambda_i})$ and consider their product probability space $(\mathscr{E}^*, \mathbf{C}^*, \nu^*)$. Also, define the discrete potential functions

$$\Phi_{\{i\}}(x^*) \equiv V(x_i^*), \qquad \Phi_{\{i,j\}}(x^*) \equiv \sum_{s \in x_i^*, t \in x_j^*} \Phi(|s - t|),$$

where z, Φ, and V are those of the continuous state space Gibbsian distribution. We define the mapping ρ which maps each continuous state space configuration x into the system of local configurations $x^* = \{x_{\Lambda_i}\}_i$. We can show that the image μ^* of μ by ρ becomes a Gibbsian distribution on $\mathscr{E}^*$ (details are omitted).

Assume that we observe data points $X_\Lambda = \{x_1, \ldots, x_n\}$ in a bounded domain Λ. Both the positions and the number of points are random. We can formulate this data using Gibbsian distributions in various ways, e.g.,

MODEL 1. We assume that there are no data points outside the domain and fit local Gibbsian distributions $\mu_\Lambda(\cdot|\varnothing)$ with empty outer configurations,

MODEL 2. We condition the first model by the number of data points and fit canonical Gibbsian distributions,

MODEL 3. We consider that the data is a part of a globally spreading configuration and fit the marginal distributions on $\mathscr{E}_\Lambda$ of a global Gibbsian distribution μ,

MODEL 4. Because marginal distributions in Model 3 have no simple likelihood expression, we consider marginal distributions $\mu_\Lambda(\cdot|y)$ conditioned by the configuration y outside the domain Λ.

From the point of view of theoretical analysis we always consider the asymptotic theory as the domain Λ expands to the whole space. In Model 2 we further have to make the assumption that the number density $n/|\Lambda|$ has a positive limit $\rho > 0$. We do not necessarily assume the existence of global Gibbsian distributions from the first in Model 1 and Model 2. This is the traditional framework in statistical physics called the "thermo-dynamical limit". In Model 4 it is implicitly assumed that we know the outer configuration y which is infinite. If the potential is of bounded range; that is, if there is a constant r_0 and $\Phi(r) = 0$ for $r > r_0$, then the conditional distribution $\mu_\Lambda(\cdot \mid y)$ actually depends only on the part of y which is within distance r_0 from Λ. The maximum approximate likelihood method of Ogata and Tanemura, introduced in §4, is based on Model 2. The Takacs-Fiksel method is based on Model 3. The maximum pseudolikelihood method of Besag is based on the fourth model. Nevertheless it is common that we have to consider Models 1 or 2 in actual applications.

3. Asymptotic efficiency of maximum likelihood estimator of potential functions

In this section we introduce the theory of asymptotic efficiency of maximum likelihood estimators (MLE) of potential functions for Model 1 from Mase [23]. The argument uses two theories. The first is the theory of LAN (Locally Asymptotically Normal) family of Le Cam. The theory of LAN families offers a unified basis of statistical asymptotic theory and has recently been applied to discussions of asymptotic inference of stochastic process models. For example, Ibragimov and Has'minski [15] and Kutoyants [20] describe asymptotic estimation theory for (1) Gaussian white noise process, (2) Diffusion process model, and (3) one-dimensional point process model from the point of view of LAN family systematically. In particular, we use the theory of ULAN (Uniformly Locally Asymptotic Normal) family formulated in Ibragimov and Has'minski [15]. The second is the (analytic) asymptotic analysis of grand partition functions given in Minlos and Pogosian [27] and Pogosian [37].

3.1. Basic assumptions. Let us fix a potential function Φ and parametrize the local energy as follows

$$V_\theta((x)_n) = nz + \alpha \sum_{1 \leq i < j \leq n} \Phi(\beta r_{ij}), \qquad \theta = (z, \alpha, \beta) \in \mathbf{R} \times (0, \infty)^2.$$

Also, we consider the two-parameter family $\theta = (z, \alpha)$ obtained by setting $\beta = 1$. Our fundamental assumptions are

(A1) $\Phi(r)$ is two-times continuously differentiable.
(B1) $\Phi(r)$ is upper semicontinuous and finite.
(A2) there are constants $c_1 > 0$, c_2, $p > 12$, and $q > 2$ such that

$$r^q \Phi(r) \to c_1, \qquad r\Phi'(r), r^2\Phi''(r) = O(\Phi(r)) \text{ for } r \to +0,$$

$$r^p \Phi(r) \to c_2, \qquad r\Phi'(r), r^2\Phi''(r) = O(|\Phi(r)|) \text{ for } r \to +\infty.$$

(B2) there are constants $c_1 > 0$, c_2, $p > 12$, and $q > 2$ such that $\lim_{r \to +0} r^q \Phi(r) = c_1$, $\lim_{r \to +\infty} r^p \Phi(r) = c_2$.

Let B and K_α be smallest constants which satisfy the following inequality;

$$V((x)_n) \geq -Bn \quad \text{for each } (x)_n \text{ (stability condition)};$$

$$|e^{-\alpha\Phi(r)} - 1| \leq K_\alpha/(1+r)^p \quad \text{for each } r > 0.$$

For each parameter $\theta = (z, \alpha, \beta)$ let $B_\theta = \alpha B$, $K_\theta = K_\alpha$ (if $\beta \geq 1$), $K_\theta = \beta^{-p}K_\alpha$ (if $\beta < 1$), $q_0(\theta) = e^{z+B_\theta+1}K_\theta$, and $q_\theta(r) = q_0(\theta)/(1+r)^{p/2}$.

(A3) Θ_0 stands for the set of those θ with both $q_0(\theta) < \frac{1}{2}$ and $\int q_\theta(|x|)\,dx < 1$. The parameter space Θ is a bounded open subset of Θ_0.

(A4) The Hessian of $p(\theta)$ is positive definite on the closure of Θ, where $p(\theta)$ is the specific free energy (the definition will be given later) for the parameter ($\theta = (z, \alpha)$).

We say a potential function is soft-core if it is finite everywhere and hard-core if $\Phi(r) = +\infty$ for $r < r_0$ for some $r_0 > 0$. In the latter case the distance between any two points cannot be smaller than r_0. In particular, two Gibbsian distributions with different hard-core distances r_0 are not mutually absolutely continuous, and there are difficulties in considering their likelihoods. Further, if hard-core distances vary with parameter values, then we are likely to have a nonregular case as is usual in statistics. This is the reason why Mase [23] discussed only the soft-core potential case. Nevertheless, if hard-core distances do not vary with parameter values (necessarily the two-parameters case), then the following discussion applies without any changes. For simplicity, for given Λ_n and θ, the corresponding Gibbsian distribution is denoted by $\mathbf{P}_{n,\theta}$, the local energy by V_θ, the grand partition function Ξ_{Λ_n} by Ξ_n, and the observed data X_{Λ_n} by X_n.

3.2. ULAN conditions. We refer the readers to the book of Ibragimov and Has'minski [15] for an explicit statement of the ULAN conditions **N1–N6** and its main consequences. The central condition **N1** is called the LAN condition, and under this condition MLE becomes a BAN estimator. That is, it has the smallest asymptotic variance among all regular estimators. If conditions **N1–N4** and **N6** hold, then MLE is locally uniformly consistent and locally uniformly asymptotic normal. If the condition **N5** holds in addition, then MLE is asymptotically efficient in the sense of the asymptotic minimax criterion for a very wide class of loss functions simultaneously.

3.3. Strong cluster estimate. The most important analytic tool to prove ULAN conditions for a family of Gibbsian distributions is the strong cluster estimate of Ursell functions due to Minlos and Pogosian. We give only an outline of the theory. See [27] and [37] for details. Let the potential function Φ satisfy the following inequality

$$|e^{-\Phi(x)} - 1| \leq K/(1+|x|)^p \quad \text{for each } x \in \mathbf{R}^2.$$

Define the constant $q_0 = e^{z+B+1}$ and the function $q(r) = q_0/(1+r)^{p/2}$. If $\int q(|x|)\,dx < 1$, then the logarithm of the grand partition function Ξ_n has the expression $\log \Xi_n = \int_{\mathscr{C}_{\Lambda_n}} \Psi(c)\,dc$. Also, the Ursell function $\Psi(c)$ has the following strong cluster estimate

$$|\Psi(c)| \leq \frac{Aq_0}{K} \sum_{\gamma \in \mathscr{L}_c} \prod_{(x,y) \in \gamma} q(|x - y|),$$

where the sum extends over the set $\mathscr{L}_c$ of all chain graphs on c, and the product extends over all vertices (x, y) of the chain γ. We associate the cluster integral

$Q(\Lambda) = \int_{\mathscr{C}_\Lambda} \Psi(c)\,dc$ with each bounded convex set Λ. Then $Q(\Lambda)$ has the following asymptotic expansion as $\Lambda \uparrow \mathbf{R}^2$

$$Q(\Lambda) = \left[\int_{\mathscr{C}} \frac{\Psi(c \cup \{0\})}{1 + \#c}\,dc\right]|\Lambda| + R(\Lambda)|\Lambda^*|,$$

where $|\Lambda|$ is the area of Λ and Λ^* is the set $\{y\colon \mathrm{dist}(x,\Lambda) < 1\}$. The residual term $R(\Lambda)$ can be bounded as $|R(\Lambda)| \leq C$ with a constant C depending only on q. The coefficient of $|\Lambda|$ on the right-hand side of this expansion is called the Gibbs' specific free energy in the context of statistical physics which corresponds to the pressure of the particle system.

3.4. Conclusion. The following is the main conclusion of this section. Let $\nabla V_\theta((x)_n)$ be the gradient of the local energy for a two- or three-parameter family.

THEOREM 1. *Let $M_{n,\theta}$ and $V_{n,\theta}$ be the mean vector and the covariance matrix of $\nabla V_\theta(X_n)$ with respect to $\mathbf{P}_{n,\theta}$, respectively. Let the matrix $\phi_n(\theta)$ and the random vector $\Delta_{n,\theta}$, which appear in the ULAN conditions, be $V_{n,\theta}^{-1/2}$ and $[\nabla V_\theta(X_n) - M_{n,\theta}]$, respectively. Under the conditions A1, A2, and A3 ULAN conditions N1–N5 hold and, in particular, MLE is a BAN estimator. For a two-parameter model the ULAN conditions N1–N6 hold under the conditions B1, B2, A3, and A4 and, in particular, MLE is asymptotically efficient.*

REMARK. Mase [22] proved the LAN condition for discrete Gibbsian distributions. Jensen [16] proved the asymptotic normality of the statistics $(\#X_\Lambda, V(X_\Lambda))$ for Models 1 and 2 using the analytic function method. Mase [25] applied his method to the proof of asymptotic equivalence of MLE's for Models 1 and 2; that is, they are asymptotic normal with the same asymptotic variance.

4. Asymptotic theory of Gibbsian distributions based on the decay of the correlation estimate

The analysis used in the preceding section is not useful for a discussion of estimators other than MLE. Recently, Jensen [17] applied the theory of decay of correlation of Gross [11], [12], Klein [18], [19], Künsch [21], and Föllmer [9] to asymptotic analysis of Gibbsian model (Model 4) and showed the asymptotic normality of MLE and the maximum pseudolikelihood estimator. The theory of Gross et al. deals with Gibbsian distributions on the lattice and guarantees a mixing-type condition under Dobrushin's uniqueness condition. The trick to applying results for discrete models to continuous models is the correspondence between continuous state space Gibbsian distribution μ and discrete Gibbsian distribution μ^* stated previously. This seems a flexible method and can be expected to become a basic tool of asymptotic analysis of Gibbsian distributions. In the following we survey this theory according to Föllmer [9] and introduce a statistical application due to Jensen.

4.1. Decay of correlation of Gibbsian distributions. We will use notations in §2. For each $i = (i_1, i_2) \in \mathbf{Z}^2$ let $\|i\| = \max\{|i_1|, |i_2|\}$. Define the distance $r = r_1 + r_2$ of configurations on a lattice Λ_i. Here r_2 is the discrete distance and the distance r_1 is defined by the following equation

$$r_1((x)_n, (y)_m) = \min_\pi \sum_{i=1}^n |x_i - y_{\pi(i)}| + m - n,$$

where the minimum on the right-hand side is taken over all permutations π, and we assume $m \geq n$. The oscillation $\delta(f)$ of a function f defined on $S_i = \mathscr{E}_{\Lambda_i}$ is $\sup |f(s) - f(t)|/r(s,t)$. The Vasserstein distance for two probability distributions μ_1, μ_2 on S_i is defined as

$$R(\mu_1, \mu_2) = \sup_f \left| \int f \, d\mu_1 - \int f \, d\mu_2 \right| / \delta(f).$$

Also, the oscillation at a coordinate i of a function f on the product space $\mathscr{E}^*$ is defined as follows

$$\delta_i(f) \equiv \sup\{|f(x^*) - f(y^*)|/r(x_i^*, y_i^*); x^* = y^* \text{ except at the coordinate } i\}.$$

Let $\mathscr{L}(\mathscr{E}^*)$ be the totality of measurable functions on $\mathscr{E}^*$ which satisfy the following condition

$$\sum_i \delta_i(f) < \infty \quad \text{and} \quad |f(x^*) - f(y^*)| < \sum_i r(x_i^*, y_i^*) \delta_i(f).$$

Suppose μ is a continuous state space Gibbsian distribution and μ^* is its discretization. We consider the following continuity assumption.

(C) For all $f \in \mathscr{L}(\mathscr{E}^*)$, the function $x^* \mapsto \int f \, d\mu^*_{\{i\}}(\cdot|x^*)$ is again a member of $\mathscr{L}(\mathscr{E}^*)$.

Dobrushin's interaction matrix $C = (C_{ik})$ is defined as follows

$$C_{ik} \equiv \sup\{R(\mu_k(\cdot|x^*), \mu_k(\cdot|y^*))/r(x_i^*, y_i^*) : x_j^* = y_j^* \text{ for each } j \neq i\}.$$

This matrix C_{ik} measures a dependency of $\mu_k(\cdot|x^*)$ on the coordinate i. With this matrix Dobrushin's uniqueness condition is expressed as

(D) $\sup_k \sum_i C_{ik} < 1$.

We also need an associated matrix $D = (D_{ik}) \equiv \sum_{n \geq 0} C^n$. From the condition (D) we can assure of the existence of D and the property

(3) $$\lim_{n \to \infty} \sum_i (C^n)_{ik} = 0 \quad \text{for all } k.$$

THEOREM 2 (Föllmer [9]). *Under condition* (3) *the family of conditional probability distributions* $\{\mu_k(\cdot|x^*)\}$ *determines* μ^*, *hence* μ, *uniquely. Further, if we have*

$$\sup_k \int r(x_k^*, y_k^*)^2 \, d\mu^*(x^*) < \infty$$

for some $y^* \in \mathscr{E}^*$, *then there is a constant depending on* σ^2 *and the following decay of correlation estimate holds*

(4) $$|\text{cov}_{\mu^*}(f, g)| \leq \sigma^2 \sum_{i,k} \delta_i(f) D_{ki} \delta_k(g) \quad \text{for all } f, g \in \mathscr{L}(\mathscr{E}^*).$$

If the probability μ is stationary, then μ^* is also stationary and we can represent C_{ik} (resp., D_{ik}) as $\widetilde{C}_{i-k}$ (resp., $\widetilde{D}_{i-k}$). Under the assumption of Theorem 2

$$\left| \sum_j \text{cov}_{\mu^*}(f, f \circ \tau_j) \right| < \infty \quad \text{if } f \in \mathscr{L}(\mathscr{E}^*).$$

Then from the relation (4), if we take measurable sets $A_i \in \sigma(X_j, j \in I_i)$ for two sets $I_1, I_2 \subset \mathbf{Z}^2$, the following strong-mixing type condition holds

$$\sup_{A_1} \sup_{A_2} |\mu^*(A_1 \cap A_2) - \mu^*(A_1)\mu^*(A_2)| \leq \sigma^2 \min\{\#I_1, \#I_2\}\alpha(d(I_1, I_2)),$$

where

$$\alpha(m) \equiv \sum_{|i| \geq m} \widetilde{D}_i, \qquad d(I_1, I_2) \equiv \inf\{|i_i - i_2|; i_j \in I_j\}.$$

From this evaluation we can apply central limit theorems of Neaderhouser [29], Bolthausen [4], or Takahata [46] under strong-mixing type conditions to sums $S(J) = \sum_{i \in J} Y_i$ of $\sigma\{X_j, \|j - i\| \leq c\}$-measurable random variables $\{Y_i\}$.

4.2. Potentials of bounded range and hard-core potentials. Jensen [17] discussed Model 4 with two-dimensional parameter $\theta = (z, \alpha)$. One class of potential functions for which we can apply the decay of the correlation estimate easily is those that are of bounded range. Assume there is a constant κ and $\Phi(r) = 0$ for all $r > \kappa$. Then $\mu_i^*(\cdot | x^*)$ does not depend on x_j^* with $\|i - j\| > \kappa$. From this fact we can show $\widetilde{C}_i = 0$ and the continuity condition (**C**) for $\|i\| > \kappa$ (Klein [18]). Using an estimate of $\sum_i \widetilde{C}_i$ due to Jensen we can be certain that the condition (**D**) is valid for sufficiently large z or sufficiently small α.

Another useful class of potential functions is the class of hard-core potential functions. Let the hard-core distance be r_0. If there is a decreasing function ψ with $|\Phi(r)| < \psi(r)$ for $r \geq r_0$ and

$$\sum_{i:\ \mathrm{dist}(\Lambda_0, \Lambda_i) \geq r_0} \psi(\mathrm{dist}(\Lambda_0, \Lambda_i)) < \infty,$$

then the continuity condition (**C**) is satisfied (Klein [18]). Jensen [17] also gave an estimate of $\sum_i \widetilde{C}_i$ in this situation, and we can prove that the condition (**D**) is valid for sufficiently large z or sufficiently small α.

4.3. Asymptotic normality of MLE and MPLE. Let X be a sample from a global Gibbsian distribution, and let X_Λ be its part which is inside Λ. The (logarithm) of pseudolikelihood of Besag is defined as follows

$$PL_\Lambda \equiv -\#X_\Lambda \cdot z - \alpha \sum_{x \in X_\Lambda} v(X\setminus\{x\}, x) - e^{-z} \int_\Lambda \exp\{-\alpha v(X, \xi)\} d\xi,$$

where we set $v(X, x) = \sum_{y \in X} \Phi(x - y)$. If the potential function is of bounded range, then PL_Λ can be computed from the part of X lying inside $\Lambda^\kappa = \{x; \mathrm{dist}(x, \Lambda) \leq \kappa\}$. Let $I_n = [-n, n]^2 \subset \mathbf{Z}^2$, and let $\Lambda_n = \bigcup_{i \in I_n} \Lambda_i$, and call the maximizer $(\hat{z}_n, \hat{\alpha}_n)$ of PL_Λ the maximum pseudolikelihood estimator (MPLE) of (z, α).

THEOREM 3 (Jensen [17]). *Assume that the potential function is either* (1) *nonnegative and of bounded range, or* (2) *hard-core, of bounded range and bounded from below. If the condition* (**D**) *holds and* $\alpha > 0$, *then the statistic* $(2n + 1)(\hat{z}_n - z, \hat{\alpha}_n - \alpha)$ *converges in law to a two-dimensional normal distribution with mean zero.* (*The explicit form of its covariance matrix can be given.*)

The logarithm of the likelihood for Model 4 is

$$L_\Lambda = -\# X_\Lambda \cdot z - \alpha V(X_\Lambda | X_{\Lambda^c}) - \log \Xi_\Lambda(X_{\Lambda^c}).$$

If the potential function is of bounded range, then L_Λ can be computed from the part of X inside Λ^κ. Let U_n (resp., J_n) be the gradient (resp., the Hessian matrix) of L_{Λ_n}, J_n is the covariance matrix of U_n with respect to $\mu_{G_n}(\cdot | X_{G_n^c})$.

THEOREM 4 (Jensen [17]). *Assume the conditions of Theorem 3. Also, assume that the eigenvalues of $(2n+1)^{-2} J_n$ are bounded from below (> 0). Let $(\hat{z}, \hat{\alpha})$ be MLE. Then the statistic $J_n^{1/2}(\hat{z}_n - z, \hat{\alpha}_n - \alpha)^T$ converges in law to a two-dimensional normal distribution with mean 0 and unit covariance matrix.*

5. Numerical procedures of estimation of potential functions from mapped point patterns

In the following we will fix a parametric family of potential functions. The parametrization need not be those introduced previously. There are (at least) four practical procedures of estimating potential functions. They are the approximated likelihood method, the pseudolikelihood method, the moment method, and the stochastic approximation method. We do not consider the stochastic approximation method here and refer the reader to the paper [28].

5.1. Approximated likelihood method. We will fix the number of points $\# X_\Lambda = n$ (Model 2). In order to estimate the parameter θ of the potential Φ_θ by MLE, we need to compute the log-likelihood

$$-\sum_{i<j} \Phi_\theta(r_{ij}) - \log \Xi_{n,\Lambda}$$

numerically. But the partition function $\Xi_{n,\Lambda}$ has no simple expression. Therefore, an approximation is indispensable. Ogata and Tanemura [30], [31] discussed how to compute this function approximately. They used analytic approximations and approximations based on computer Monte Carlo simulations. Analytic approximations are based on so-called low-density expansions. The virial expansion and the cluster expansion are well known in statistical physics. For example, the virial expansion yields the following approximation

$$n^{-1} \log \Xi_{n,\Lambda} \approx (b/2) \int_{\mathbf{R}^2} f_{12} \, dx_2 + (b^2/4) \int_{\mathbf{R}^4} f_{12} f_{13} f_{23} \, dx_2 dx_3 + \cdots, \qquad b = n/|\Lambda|,$$

where $f_{ij} = \exp\{-\Phi_\theta(r_{ij})\} - 1$, see Mase [24] for details. This approximation is always possible if we make multidimensional numerical integrals. But higher order terms consist of sums of a large number of integrals with fairly complicated integrands and, in practice, expansions up to the seventh or eighth term are possible at most, see, for example, Hoover and DeRocco [14]. This limitation shows that it is difficult to use this approximation if strong interactions are present. Ogata and Tanemura [32], [33] used this approximation to estimate interactions for spatially nonstationary point processes or marked point processes.

Another type of approximation is based on a Monte-Carlo computer experiment. This can yield approximations effective even for a strong interaction if the potential function is nonnegative and of the form $\Phi_\theta(r) = \Phi(r/\theta)$, see Ogata and Tanemura

FIGURE 1(a). A pattern of steel balls with diameter 0.5mm in a plastic container. They were electronically charged by shaking them fiercely.

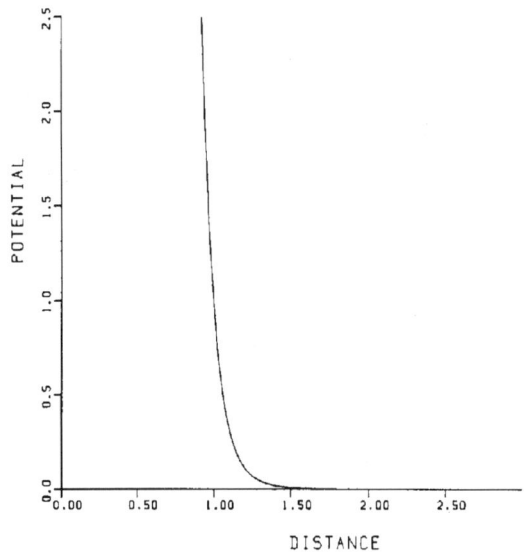

FIGURE 1(b). MLE of the corresponding potential function. $\Phi_1(r) = (\theta/r)^n$; $\hat{n} = 5.79 \pm 0.66$; $\hat{\psi} = 1.64$; $\hat{\tau} = 0.42$.

[31], [35]. Let us define the reduced density of a configuration by $\tau = n\theta^2/|\Lambda|$. Using τ as a variable we can express the logarithm of the partition function as follows

$$\log \Xi_{n,\Lambda} = -n \int_0^\tau \psi(t)/t \, dt,$$

where ψ is the function called the compressibility factor. In order to evaluate numerical values of ψ corresponding to a given potential function, we simulate point patterns on the unit torus $|\Lambda| = [0,1] \times [0,1]$ for various values of τ_i. We refer readers to Metropolis et al. [26], Wood [47], Ripley [40], or Ogata and Tanemura [30] for asymptotic computer generation of Gibbsian point patterns. From these simulated patterns we estimate values and errors of $\psi(\tau_i)$ by the following weighted mean of the derivative $\Phi'(\cdot)$ of the potential function,

$$\hat{\psi}(\tau_i) = -\frac{1}{2n} \frac{1}{T} \sum_{t=1}^{T} \sum_{i<j} \frac{r_{ij}}{\sqrt{\tau_i}} \cdot \Phi'\left(\frac{r_{ij}}{\sqrt{\tau_i}}\right).$$

Finally we interpolate them by rational functions $(p_1\tau + p_2\tau^2 + p_3\tau^3)/(1 + q_1\tau + q_2\tau^2)$ (Padé approximation). Various examples of procedures of this section are given in Ogata and Tanemura [35]. See Figure 1 for an equilibrium pattern of steel balls and its analysis.

5.2. Pseudolikelihood method. Besag [1] proposed the pseudolikelihood method in order to avoid difficulties in calculating likelihoods of Markov random fields on lattice. Also, he showed that this method is also applicable to a continuous state space model if one performs a limiting procedure of shrinking lattice size, see Besag [2], Besag et al. [3].

If we are interested only in the estimation of θ, Besag's argument can be simplified as follows. We let $X = X_\Lambda = \{x_i\}$ and consider Model 2. The conditional intensity function at a point $x \in \Lambda$ under the condition that the configuration X^c in $\Lambda \setminus \{x\}$ is given becomes, by Definition 1,

$$e^{-z} \exp\left\{-\sum_j \Phi_\theta(|x - x_j|)\right\} \equiv e^{-z} \gamma_0(x|X^C).$$

Then the conditional probability density function at each x_i is

$$\gamma_0(x_i|X^i) / \int_\Lambda \gamma_0(x|X) \, dx.$$

Here X^i is the configuration X excluding the point x_i. The logarithm of the formal product of them for each i formally is

(5) $$\sum_{i=1}^{n} \log \gamma_0(x_i|X^i) - n \log \int_\Lambda \gamma_0(x|X) \, dx,$$

and the maximizer θ of this expression is equal to that of MPLE of Besag, see §4.3. In the study of Diggle et al. [6] the integral in (5) was calculated numerically using the 40×40 square lattice. It seems there is no serious problem of numerical precision if we analyze point patterns with point numbers about $n = 100$.

5.3. Moment (Takacs-Fiksel) method. A great advantage of considering Model 3, that is, a global Gibbsian distribution, is the validity of the following Palm-Khinchine theorem, see Stoyan et al. [44],

(6) $$\lambda \mathbf{E}_0[Z] = \mathbf{E}\left[Z \cdot \exp\left\{-z - \sum_i \Phi_\theta(|x_i|)\right\}\right],$$

where the mean on the right-hand side is taken with respect to the global Gibbsian distribution, that on the left-hand side is taken with respect to the corresponding Palm distribution (the conditional distribution of configurations other than the point at the origin under the condition that configurations have a point at the origin), Z is an arbitrary function of the data points X, and λ is the mean number of points per unit area (intensity). Takacs [45] proposed an idea of replacing both sides of equation (6) by corresponding sampling equivalents and solving the resulting equation with respect to θ. Fiksel [7], [8] further extended this idea to both models with general potentials and marked point process models. More precisely, one substitutes certain test functions $Z_1, \ldots, Z_m$ into the equation (6) and obtains sample estimates $\hat{L}_k(z, \theta)$ and $\widehat{R}_k(z, \theta)$ of both sides. Finally, we seek the minimizer of the expression

$$S(z, \theta) = \sum_{k=1}^{m} \{\hat{L}_k(z, \theta) - \widehat{R}_k(z, \theta)\}^2.$$

Of course, the quality of the resulting estimators crucially depends on the choice of test functions. Takacs and Fiksel used the following test functions

$$Z_k = N(r_k) \exp\left\{z + \sum_{i=1}^{n} \Phi_\theta(|x_i|)\right\},$$

where $N(r)$ is the number of points inside the disk with radius r and center at origin. The merit of this choice is that we can estimate the right-hand side of (6) simply by $R_k(z, \theta) = \lambda \pi r_k^2$. Nevertheless, in our experiment, we had better results with the choice $Z_k = N(r_k)$.

6. A simulation comparison of various potential estimators

We made a simulation study to compare the performance of the preceding three estimators. We used several typical repulsive potential functions and considered three boundary conditions, that is, the free boundary, the periodical boundary, and the weighted boundary. The characteristics of each estimator seen from this experiment are summarized as follows.

(1) MLE calculated from the approximated likelihood based on low density expansions were most effective if the interactions were weak. If not, they caused a remarkable bias. Since the standard deviation is smallest it is suitable for the likelihood ratio test of the null hypotheses of Poisson processes. The approximated likelihood, using the virial expansion up to the second or the third term, can be easily calculated for every potential function.

(2) MLE calculated from the approximated likelihood based on Monte Carlo simulations was unbiased and most effective under the periodic boundary condition for every degree of interaction. For other boundary conditions it gave biased estimates if the interactions were large. However, this may be a mere consequence of the fact that we could not generate equilibrium patterns with strong interactions except by using the periodic boundary. Since we must make a Monte Carlo simulation for each potential function this method is very laborious. Nevertheless, we can get the maximum value of likelihoods which is necessary for model selections using, e.g., the AIC method.

(3) MPLE were usually robust and reasonable for all boundary conditions. However, it showed remarkable bias for strong interactions even for the periodic boundary. Pseudolikelihoods can be easily computed using numerical integration.

(4) The moment method usually gave unbiased and reasonable estimators under the periodic boundary condition even for strong interactions. However, the fluctuation of estimators became larger for weak interactions. Sometimes we had outlying values and had to be careful. Its practical merit is that we can estimate any potential function easily. There is the possibility of lessening these fluctuations if we choose better test functions.

7. Conclusion

We have given an overview of the present condition of the theory and practice of statistical estimation of mapped point pattern data via Gibbsian models. As the readers have surely noticed, we are making trial and error research, and there is still a big gap between the theory and the practice. We hope this brief exposition will arouse our reader's interest in this new and challenging field.

References

1. J. Besag, *Spatial interaction and the statistical analysis for spatial data*, J. Roy. Statist. Soc. **B 36** (1974), 192–236.
2. _____, *Some methods of statistical analysis for spatial data*, Bull. Inst. Internat. Statist. **47** (1978), 77–92.
3. J. Besag, R. Milne, and S. Zachary, *Point process limits of lattice processes*, J. Appl. Probab. **19** (1982), 210–216.
4. E. Bolthausen, *On the central limit theorem for stationary mixing random fields*, Ann. Probab. **10** (1982), 1047–1050.
5. P. J. Diggle, *Statistical analysis of spatial point patterns*, Academic Press, London, 1983.
6. P. J. Diggle, T. Fiksel, P. Grabarnik, Y. Ogata, D. Stoyan, and M. Tanemura, *On parameter estimation for pairwise interaction point processes*, Internat. Statist. Rev. **62** (1994), 99–117.
7. T. Fiksel, *Estimation of parameterized pair potentials of marked and nonmarked Gibbsian point processes*, Elektron. Inform. Kybernet. **20** (1984), 270–278.
8. _____, *Estimation of interaction potentials of Gibbsian point processes*, Math. Operationsforsch. Statist. Ser. Statist. **19** (1988), 77–86.
9. H. Föllmer, *A covariance estimate for Gibbs measures*, J. Funct. Anal. **46** (1982), 387–395.
10. H.-O. Georgii, *Gibbs measures and phase transitions*, De Gruyter, Berlin, 1988.
11. L. Gross, *Decay of correlations in classical lattice models at high temperature*, Comm. Math. Phys. **68** (1979), 9–27.
12. _____, *Absence of second-order phase transitions in the Dobrushin uniqueness region*, J. Statist. Phys. **25** (1981), 57–72.
13. K. H. Hanisch and D. Stoyan, *Remarks on statistical inference and prediction for a hard-core clustering model*, Math. Operationsforsch. Statist. Ser. Statist. **14** (1983), 559–567.
14. W. G. Hoover and A. G. DeRocco, *Sixth and seventh virial coefficients for the parallel hard-cube model*, J. Chem. Phys. **36** (1962), 3141–3162.
15. I. A. Ibragimov and R. Z. Has′minski, *Statistical estimation, asymptotic theory*, Springer, Berlin, Heidelberg, New York, 1981.
16. J. L. Jensen, *A note on asymptotic normality in the thermodynamical limit at low densities*, Adv. in Appl. Math. **12** (1991), 387–399.
17. _____, *Asymptotic normality of estimates in spatial point processes*, Research Reports No. 210, Department of Theoretical Statistics, Aarhus University, 1990.
18. D. Klein, *Dobrushin uniqueness techniques and the decay of correlations in continuum statistical mechanics*, Comm. Math. Phys. **86** (1982), 227–246.
19. _____, *Convergence of grand canonical Gibbs measures*, Comm. Math. Phys. **92** (1984), 295–308.
20. Yu. A. Kutoyants, *Parameter estimation for stochastic processes*, Heldermann, Berlin, 1984.

21. H. Künsch, *Decay of correlation under Dobrushin's uniqueness condition and its applications*, Comm. Math. Phys. **84** (1982), 207–222.
22. S. Mase, *Locally asymptotic normality of Gibbs models on a lattice*, J. Appl. Probab. **16** (1984), 585–602.
23. _____, *Uniform LAN condition of planar Gibbsian point processes and optimality of maximum likelihood estimators of soft-core potential functions*, Probab. Theory Related Fields **92** (1992), 51–67.
24. _____, *Mean characteristics of Gibbsian point processes*, Ann. Inst. Statist. Math. **42** (1990), 203–220.
25. _____, *Asymptotic equivalence of grand canonical MLE and canonical MLE of pair potential functions of Gibbsian point process model*, Technical Report Series No. 292, Hiroshima Statistical Research Group, 1991.
26. N. Metropolis, A. W. Rosenbluth, M. N. Rosenbluth, A. H. Teller, and E. Teller, *Equations of state calculations by fast computing machines*, J. Chem. Phys. **21** (1953), 1087–1092.
27. R. A. Minlos and S. K. Pogosian, *Estimates of Ursell functions, group functions, and their derivatives*, Theoret. and Math. Phys. **31** (1977), 408–418.
28. R. A. Moyeed and A. J. Baddeley, *Stochastic approximation of the MLE for a spatial point pattern*, Scand. J. Statist. **18** (1991), 39–50.
29. C. C. Neaderhouser, *Some limit theorems for random fields*, Comm. Math. Phys. **61** (1978), 293–305.
30. Y. Ogata and M. Tanemura, *Estimation of interaction potentials of spatial point patterns through the maximum likelihood procedure*, Ann. Inst. Statist. Math. **33 B** (1981), 315–338.
31. _____, *Likelihood analysis of spatial point patterns*, J. Roy. Statist. Soc. **B 46** (1984), 496–518.
32. _____, *Estimation of interaction potentials of marked spatial point processes through the maximum likelihood method*, Biometrics **41** (1985), 421–433.
33. _____, *Likelihood estimation of interaction potentials and external fields of inhomogeneous spatial point patterns*, Proceedings of Pacific Statistical Congress (I. S. Francis, B. F. J. Manly, and F. C. Lam, eds.), North-Holland, Amsterdam, 1986, pp. 150–154.
34. _____, *Likelihood analysis using Gibbsian models*, Tokei Suuri **35** (1987), 257–275. (Japanese)
35. _____, *Likelihood estimation of soft-core interaction potentials for Gibbsian point patterns*, Ann. Inst. Statist. Math. **41** (1989), 583–600.
36. A. Penttinen, *Modelling interactions in spatial point patterns: parameter estimation by the maximum-likelihood method*, Jyväskyla Studies in Comput. Sci., Econom. and Statist. **7** (1984), 1–107.
37. S. K. Pogosian, *Asymptotic expansion of the logarithm of the partition function*, Comm. Math. Phys. **95** (1984), 227–245.
38. C. Preston, *Random fields*, Springer, Berlin, Heidelberg, and New York, 1976.
39. B. D. Ripley, *Simulating spatial patterns: dependent samples from a multivariate density*, Appl. Statist. **28** (1979), 109–112.
40. _____, *Statistical inference for spatial processes*, Cambridge Univ. Press, Cambridge, 1988.
41. D. Ruelle, *Statistical mechanics: Rigorous results*, Benjamin, New York, 1969.
42. D. Stoyan, W. S. Kendall, and J. Mecke, *Stochastic geometry and its applications*, Wiley, New York, 1987.
43. R. Takacs, *Estimator for the pair-potential of a Gibbsian point process*, Math. Operationsforsch. Statist. Ser. Statist. **17** (1986), 429–433.
44. H. Takahata, *On the central limit theorem for weakly dependent random fields*, Yokohama Math. J. **31** (1983), 67–77.
45. W. W. Wood, *Monte Carlo studies of simple liquids*, Physics of Simple Liquids (H. N. V. Temperley, J. S. Rowlinson, and G. S. Rushbrook, eds.), North-Holland, Amsterdam, 1968, pp. 115–230.

Translated by SHIGERU MASE

Spectra in Random Media

Shin Ozawa

1. Introduction

If a small amount of metal B is contained in metal A, then what can one say about the physical properties of that medium? Here by a physical property we mean, for example, electric conductivity. It is easy to see that electric conductivity is determined by a configuration of metal B. If we consider random media, then we fix a probability space on the configuration of the metal B and we consider statistics of the physical property. What is the effective conductivity? What is the most realizable state? This is an old problem. We can find it in J. C. Maxwell: *A treatise on electricity and magnetism*, Vol. 1, §314, Clarendon Press, Oxford, 1881. The problem has been studied by many authors. However, we can say that there is yet no satisfactory theory that is mathematically rigorous.

On the other hand, there is the problem of homogenization. In this problem the included metal is included periodically and the size of the mesh is a parameter of the configuration of the included metal. What can we say about the asymptotic properties of the metal when the size of mesh tends to zero? The homogenization problem, as a problem in mathematics, has yielded interesting results such as those in [3], [4], [10], and [32]. However, to examine random media we need essentially different mathematical methods.

In this note we do not consider random media with random potential as is done in [6], [9]. Instead, we consider random media with obstacles. We do not touch upon homogenization with randomness that is examined in [13], for example.

2. Recent results

In this section we consider recent topics in the research on the spectra of random media.

Let Ω be a bounded domain in $\mathbb{R}^2$ with smooth boundary $\partial\Omega$. Let w be a point of Ω and $B(\varepsilon; w)$ the disk of radius ε with center w. We set $\Omega_\varepsilon = \Omega \setminus \overline{B(\varepsilon; w)}$ and

---

1991 *Mathematics Subject Classification*. Primary 35B20.
This article originally appeared in Japanese in Sûgaku **44** (4) (1992), 306–319.

set $\Omega_0 = \Omega$. Consider the following eigenvalue problem

$$-\Delta u(x) = \lambda u(x), \quad x \in \Omega_\varepsilon,$$

$$u(x) = 0, \quad x \in \partial\Omega,$$

$$u(x) + k\varepsilon^\sigma \frac{\partial}{\partial v_x} u(x) = 0, \quad x \in \partial B(\varepsilon; w).$$

Here k is a positive constant, ε is a parameter, σ is a fixed real number, $-\Delta =$ minus Laplacian, $\partial/\partial v_x$ is a derivative along the exterior normal direction at the boundary. Let $\lambda_j(\varepsilon) > 0$ be the jth eigenvalue of the above problem and $\varphi_j(x)$ be the jth L^2-normalized eigenfunction of $-\Delta$ in Ω under the Dirichlet condition on $\partial\Omega$.

We have the following theorem.

THEOREM 2.1. *Assume that $\lambda_j(0)$ is a simple eigenvalue. Then,*

(2.1) $$\lambda_j(\varepsilon) - \lambda_j(0) = 2\pi\varepsilon^{1-\sigma} k^{-1} \varphi_j(w)^2 + O(\varepsilon^{2-2\sigma}(\log \varepsilon)^2)$$

for $\sigma \in (0, 1)$ and

(2.2) $$\lambda_j(\varepsilon) - \lambda_j(0) = 2\pi\varepsilon k^{-1} \varphi_j(w)^2 + O(\varepsilon^{1+\beta})$$

for $\sigma = 0$ and for any $\beta \in (0, 1)$.

This result is in [17]. A more general formula for $\lambda_j(\varepsilon) - \lambda_j(0)$ for $\sigma \in \mathbb{R}$ is analyzed in [21]. It depends on σ in an interesting manner.

Here we recall the importance of the eigenvalue of the Laplacian. In the heat equation and the wave equation the eigenvalue plays a decisive role.

Note the term $\varepsilon^{1-\sigma}$ in the formula (2.1). We can say that the shift of the eigenvalue is of order $(\alpha/N)^{1-\sigma}$ when $\varepsilon = \alpha/N$.

Let $\omega(N)$ be a set of points $(\omega_1, \ldots, \omega_n)$ where $n = [N^{1-\sigma}]$. We set $\Omega_{w(N)} = \Omega \setminus \overline{\bigcup_{j=1}^n B(\alpha/N, w_j)}$ and assume that $B(\alpha/N, w_i) \cap B(\alpha/N, w_j) = \emptyset$ for $i \neq j$.

We consider the following eigenvalue problem.

$$-\Delta u(x) = \lambda u(x), \quad x \in \Omega_{w(N)},$$

$$u(x) = 0, \quad x \in \partial\Omega,$$

$$u(x) + k\varepsilon^\sigma \frac{\partial}{\partial v_x} u(x) = 0, \quad x \in \partial B(\alpha/N; w_i) \ (i = 1, \ldots, n).$$

Let $\lambda_j(w(N))$ be the jth eigenvalue of the above problem. We want to find asymptotic properties of $\lambda_j(w(N))$ when $N \to \infty$. If $(w_1, \ldots, w_n)$ is randomly distributed, then finding the asymptotics of $\lambda_j(w(N))$ is an interesting problem.

We must fix a probability space in order to consider randomness. Let $V(x) > 0$ be a continuous function on $\overline{\Omega}$ satisfying $\int_\Omega V(x)\,dx = 1$. Then we set

$$P(x \in A) = \int_A V(x)\,dx.$$

Thus, Ω can be thought of as a probability space with this statistical law. Let Ω^n be the product probability space. This corresponds to the fact that each of the centers of the balls $w_1, \ldots, w_n$ is independently moving on Ω. This fact contradicts the fact that $B(\alpha/N; w_i) \cap B(\alpha/N; w_j) \neq \emptyset$. However, if we set $\sigma \in (0,1)$, then we have

$$\lim_{N \to \infty} P(B(\alpha/N; w_i) \cap B(\alpha/N; w_j) = \emptyset) = 1.$$

Therefore, we do not need to study the case where $B(\alpha/N; w_i) \cap B(\alpha/N; w_j) \neq \emptyset$. The following result holds.

THEOREM 2.2. *Fix j. Fix $\sigma \in (1/3, 1)$. Then there exists a small $c_0 > 0$ such that for any $\alpha \in (0, c_0)$ we have*

$$\lim_{n \to \infty} P(w(N) \in \Omega^n; N^\rho |\lambda_j(w(N)) - \lambda_j^V| < \varepsilon) = 1$$

for any fixed $\varepsilon > 0$. Here ρ is a number satisfying $\rho \in (0, (1-\sigma)/2)$, and μ_j^V is the jth eigenvalue of the Schrödinger operator $-\Delta + 2\pi k^{-1}\alpha^{1-\sigma} V(x)$ in Ω under the Dirichlet condition.

Next we give a heuristic discussion. First, we should remark that the total area of the disks $\pi(\alpha/N)^2[N^{1-\sigma}]$ tends to 0 if N tends to ∞. However, the eigenvalue is shifted from the eigenvalue of $-\Delta$ to the eigenvalue of $-\Delta + 2\pi k^{-1}\alpha^{1-\sigma} V(x)$. This shift is a version of the Lenz shift that is stated later. Next we explain why this curious phenomenon occurs. From (2.1) it is natural to imagine that

$$(2.2) \quad \lambda_j((n \text{ disks})^C) = \lambda_j(0) + \sum_{k=1}^n 2\pi(\alpha/N)^{1-\sigma} k^{-1} \varphi_j(w_k)^2 + R(w_1, \ldots, w_n).$$

This formula is correct, since we write the remainder as R. Next we recall the law of large numbers for the sum of independent, identically distributed random variables.

Let $\{X_i\}_{i=1}^\infty$ be i.i.d. random variable with mean E. Then by the law of large numbers, we have

$$m^{-1} \sum_{i=1}^m X_i \to E(X_i) = E.$$

Now we apply this formula for $V(x) \equiv |\Omega|^{-1}$ to (2.2). Then we see that the second term in the right-hand side of (2.2) tends to $2\pi\alpha^{1-\sigma} k^{-1} |\Omega|^{-1}$ in probability. If we can get that $R(w_1, \ldots, w_n)$ tends to 0, then it follows that $\lambda_j(0)$ shifts to $\lambda_j(0) + 2\pi\alpha^{1-\sigma} k^{-1} |\Omega|^{-1}$.

However, this heuristic explanation does not hold when $V \not\equiv |\Omega|^{-1}$. Since the eigenfunction is subject to the bias of an averaged density of obstacles, the eigenfunction must be shifted to another one. We believe that Theorem 2.2 can be generalized to $\sigma \in [0, 1)$ and that the restriction for the size of α can be removed.

3. Mark Kac

The author was fascinated by the interesting paper Kac [8]. Almost all of this paper is a heuristic consideration and there are no rigorous proofs. This paper treats three topics. One of them is the Lenz shift (its name appeared in §2).

Let Ω be a bounded domain in $\mathbb{R}^3$ with smooth boundary $\partial \Omega$. Let $m = 1, 2, \ldots$ be a parameter. We remove m balls of radius α/m whose centers are $w_1, \ldots, w_m$. That is, $\Omega_{w(m)} = \Omega \setminus \overline{m\text{-balls}}$, $w(m) = (w_1, \ldots, w_m)$. Let $\mu_j(w(m))$ be the jth eigenvalue of the following:

$$-\Delta u(x) = \lambda u(x), \quad x \in \Omega_{w(m)},$$
$$u(x) = 0, \quad x \in \partial \Omega_{w(m)}.$$

We consider this problem statistically as stated above. Let $V(x)$ be as before. Let Ω be a probability space, and let Ω^m be its product probability space. We set $V(x) = |\Omega|^{-1}$. Assume that μ_j is simple. Then

(3.1) $$\lim_{m \to \infty} P(w(m) \in \Omega^m; |\mu_j(w(m)) - (\mu_j + 4\pi\alpha|\Omega|^{-1})| < \varepsilon) = 1.$$

Here μ_j is the jth eigenvalue of $-\Delta$ in Ω under the Dirichlet condition. Rauch-Taylor [23] proved the generalization of (3.1). When $V(x) \not\equiv |\Omega|^{-1}$, then (3.1) holds if we replace $\mu_j + 4\pi\alpha|\Omega|^{-1}$ by the jth eigenvalue of the Schrödinger operator $-\Delta + 4\pi\alpha V(x)$. Kac called $4\pi\alpha|\Omega|^{-1}$ the Lenz shift. We do not know why he used the name Lenz.

His method of proof is stimulating, and it has great influence on probability theory. Let $r(\tau)$; $0 \leq \tau \leq t$ be the path of a Brownian motion. We set $W_\delta(t) = \{x \in \mathbb{R}^3;$ $|x - r(\tau)| < \delta$ for some $\tau\}$. $W_\delta(t)$ is called a Wiener sausage by Marc Kac. His idea is as follows. Since the boundary of $B(\alpha/m; w_i)$ is an absorbing wall and the path is killed when it touches $\partial B(\alpha/m; w_i)$, we have the equivalence:

$W_{\alpha/m}(t)$ does not include $w_1, \ldots, w_m$

$\Leftrightarrow$ the Brownian path $r(\tau)$ does not touch m-balls.

From this equivalence and the standard theory of Brownian motion together with some heuristics about the Wiener sausage, he proved (3.1). Since analysis of Brownian motion $\approx$ analysis of the heat equation, he proved his result using a probabilistic representation of

$$\sum_{j=1}^{\infty} e^{-\mu_j(w(m))t}.$$

An exact proof is in Simon [25, pp. 231–245]. It should be noted that [8] includes a conjecture that has been solved by Donsker-Varadhan [33] using a large deviation theory. M. Kac's influence is very great. Kac [8] also studied the probabilistic explanation of scattering length. See Taylor [30] and Takahashi [29].

The author is stimulated by the part on page 525 of Kac [8]. He expressed a negative opinion of an analytic method and an especially strong negative opinion of perturbative calculus. His explanation has incorrect points. In the same year there appeared Huruslov-Marchenko [7], which studied a somewhat more general problem by an analytic method. We have studied his theorem using perturbation theory, as is presented in the next section. Huruslov-Marchenko [7] is potential theoretic and is not perturbative calculus. It is a very good book.

4. Perturbation theory

Here perturbation theory means the singular variation of domain. Let us recall Theorem 2.1. Let $G_\varepsilon(x, y)$ be the Green function of $-\Delta$ in Ω_ε under the boundary condition associated with $\lambda_j(w(m))$. A key to proving Theorem 2.1 is to approximate

$G_\varepsilon(x, y)$ by the integral kernel constructed by the Green function $G(x, y)$ of $-\Delta$ in Ω under the Dirichlet condition. See [18].

Now we want to explain our result in [16] that generalizes Kac's theorem. Let $m = 1, 2, 3, \ldots$ be a parameter. Fix $\beta \in [1, 3)$. Let α be a constant. We consider $n = [m^\beta]$ points $w_1, \ldots, w_n$. Let Ω be a bounded domain in $\mathbb{R}^3$ with smooth boundary $\partial\Omega$. We remove n balls of radius α/m whose centers are $w_1, \ldots, w_n$ and we get $\Omega_{w(m)} = \Omega \setminus \overline{n\text{-balls}}$. Let $\mu_j(w(m))$ be the jth eigenvalue of $-\Delta$ in $\Omega_{w(m)}$ under the Dirichlet condition on $\partial\Omega_{w(m)}$. If $\Omega_{w(m)}$ is not connected, let Ω^0 be a connected component. Let $\mu_k(\Omega^0)$ be the kth eigenvalue of $-\Delta$ in Ω^0 under the Dirichlet condition on $\partial\Omega^0$. We line up all of the (Ω^0, kth eigenvalue) in an increasing order, and we get the jth eigenvalue. We also consider the problem statistically. Let $V(x)$ be as stated above. We consider Ω as a probability space, and let Ω^n be its product probability space.

We have the following Theorem 4.1. It should be noted that the total volume of the balls tends to zero for $\beta \in [1, 3)$.

THEOREM 4.1 ([16]). *Fix $\beta \in [1, 5/4)$, $V(x) \equiv |\Omega|^{-1}$. Assume that the jth eigenvalue of $-\Delta$ in Ω under the Dirichlet condition on $\partial\Omega$ is simple. Then*

$$\lim_{m \to \infty} P(w(m) \in \Omega^n; m^\rho |\mu_j(w(m)) - (\mu_j + 4\pi\alpha m^{\beta-1} |\Omega|^{-1})| < \varepsilon) = 1$$

for any $\rho < 1 - (\beta/2)$. Moreover,

$$m^{1-(\beta/2)}(\mu_j(w(m)) - (\mu_j + 4\pi\alpha m^{\beta-1} |\Omega|^{-1}))$$

tends in distribution to a Gaussian random variable Π_j of mean $E(\pi_j) = 0$ and variance $E(\Pi_j^2) = (4\pi\alpha)^2 (\int_\Omega \varphi_j(x)^4 \, dx - |\Omega|^{-1})|\Omega|^{-1}$ for $m \to \infty$.

Note that eigenvalue is of order $m^{\beta-1}$ if $\beta > 1$. Much of the last half of the theorem is due to Figari-Orlandi-Teta [5]. They examined the case $\beta = 1$ and considered the fluctuation. It is an important article in spite of a gap in the proof.

We cannot say anything in the case $\beta \geq 5/4$. Thus, the case $\beta \geq 5/4$ is an open problem. Here we would like to emphasize Theorem 4.2, below, which is an effective theorem for any $\beta \in [1, 3)$ concerning the approximation of the resolvent. It is a strong result.

It is easy to see that the following holds.

(4.1) If we divide $\Omega_{w(m)}$ into the disjoint union of its components, then all but one of these components have the property that their diameter is smaller than $m^{-1}(\log m)^2$.

Then we can say that

$$\lim_{m \to \infty} P(w(m) \in \Omega^m; (4.1)) = 1.$$

Therefore, only one component ω of $\Omega_{w(m)}$ is big. Since the order of the eigenvalue is $m^{\beta-1}$ we take $(-\Delta + Tm^{\beta-1})^{-1}$ as the resolvent.

THEOREM 4.2. *Fix* $\beta \in [1,3)$. *Let* $G_{w(m)}$ *be the integral kernel of* $(-\Delta + Tm^{\beta-1})^{-1}$ *in* ω *under the Dirichlet condition on* $\partial\omega$. *Here* T *is a sufficiently large fixed number. We fix* $\varepsilon > 0$. *Let* χ_ω *be the characteristic function of* ω, *and let*

$$A = (-\Delta + 4\pi\alpha V(x)m^{\beta-1} + Tm^{\beta-1})^{-1}$$

be the integral kernel in Ω *under the Dirichlet condition on* $\partial\Omega$. *Then the probability that*

(4.2) $\qquad \|(\mathbb{G}_{w(m)}\chi_\omega - \chi_\omega A\chi_\omega)f\|_{L^2(\omega)} \leq m^{6\varepsilon} m^{-(3/4)\beta+(1/4)} \|f\|_{L^\infty(\Omega)}$

holds for any $f \in L^\infty(\Omega)$ *is greater than* $1 - m^{-\varepsilon}$.

We know the resolvent equation

$$(B+\lambda)^{-1} - (B'+\lambda)^{-1} = -(B+\lambda)^{-1}(B-B')(B'+\lambda)^{-1}.$$

Thus, from Theorem 4.2 we have that

$$-\Delta_{|w} = -\Delta + 4\pi\alpha m^{\beta-1}V + o(m^{\beta-1})$$

in some sense. Theorem 4.1 is a by-product of analyzing Theorem 4.2.

As a result, perturbation theory gives the analytic result which includes the central limit theorem.

5. Perturbation theory. II

We frequently encounter Green's function method in physics. This is a standard method in physics. However, in quantum physics, we cannot find any mathematically rigorous treatment of Green's function method for the many body problems. Zaiman [31] is a good book to read about Green's function method. It explains how many physical laws are derived using Green's function. We believe that the construction of a mathematical rigorous theory of the Green's function method is a big theme in mathematical physics. For example, Fröhlich-Spencer [6] is an excellent paper on mathematical physics which derives a mathematically rigorous result by using Green's function method.

In our research, the main point of view in [15] is point interaction approximation. We would like to examine this as we explain Theorem 4.2.

Let $G(x,y;w(m))$ be the integral kernel on ω under the Dirichlet condition on $\partial\omega$ of $(-\Delta + Tm^{\beta-1})^{-1}$. A direct analysis is not easy. We approximate $G(x,y;w(m))$ by using $G(x,y) = (-\Delta + Tm^{\beta-1})^{-1}(x,y)$ in Ω under the Dirichlet condition on $\partial\Omega$.

Let $w_1, \ldots, w_n$ be the centers of the balls. We introduce the following integral kernel

(5.1)
$$h(x,y;w(m)) = G(x,y) - (4\pi\alpha/m)\sum_{i=1}^{n} G(x,w_i)G(w_i,y)$$
$$+ \sum_{s=2}^{m^*} (-4\pi\alpha/m)^s \sum_{(s)} G(x,w_{i_1})G_{I(s)}G(w_{i_s},y).$$

Here $G_{I(s)} = G(w_{i_1}, w_{i_2}) \cdots G(w_{i_{s-1}}, w_{i_s})$, where the indices in $\sum_{(s)}$ run over all $1 \leq i_1, \ldots, i_s \leq n$ such that $i_\nu \neq i_\mu$ if $\nu \neq \mu$, $m^* = (\log m)^2$. The sum $\sum_{(s)}$ is called a self-avoiding sum. Thus, $G(w_{i_1}, w_{i_2})G(w_{i_2}, w_{i_1})$ is not contained in the sum.

If the sum of indices is not self-avoiding, then the singularity of the Green function grows stronger beyond our control. Treating the integral kernel by moderating the singularity of the integral kernel is one of the author's themes.

We set

$$\mathbb{G}_{w(m)}f(x) = \int_\omega G(x,y;w(m))f(y)\,dy,$$

$$\mathbb{H}_{w(m)}f(x) = \int_\omega h(x,y;w(m))f(y)\,dy.$$

Then we get the following

PROPOSITION 5.1. *For any fixed $\varepsilon > 0$, we have*

$$P\big(w(m) \in \Omega^n; \|\mathbb{H}_{w(m)} - \mathbb{G}_{w(m)}\|_{\mathscr{L}(L^2(\omega),L^2(\omega))} \leq m^{5\varepsilon}m^{(\beta-4)/5}\big) \geq 1 - m^{-\varepsilon}.$$

Here $\mathscr{L}(L^2(\omega), L^2(\omega))$ denotes the operator norm on $L^2(\omega)$. For results on $\mathscr{L}(L^\infty(\omega), L^2(\omega))$, see [16].

Next we explain our reason for considering $h(x,y;w(m))$. In the book Schiffer-Spencer [34], on Riemann surfaces, the authors proposed a variational formula for Green's function under singular variations of the domain. Let us consider a bounded domain $\Omega \subset \mathbb{R}^2$. Fix $w \in \Omega$. Let $B(\varepsilon;w)$ be an open disk of radius ε with center w. We set $\Omega_\varepsilon = \Omega \setminus \overline{B(\varepsilon;w)}$. Let $G_\varepsilon(x,y), (G(x,y),$ respectively) be Green's function of the Laplacians in Ω_ε (Ω, respectively) under the Dirichlet condition on $\partial\Omega_\varepsilon$, ($\partial\Omega$, respectively). Then it can be shown that

$$G_\varepsilon(x,y) = G(x,y) + (2\pi)(\log\varepsilon)^{-1}G(x,w)G(w,y) + O(\varepsilon).$$

The remainder $O(\varepsilon)$ is a pointwise remainder.

If we replace $\mathbb{R}^2$ by $\mathbb{R}^3$, then we have

$$G_\varepsilon(x,y) = G(x,y) - 4\pi\varepsilon G(x,w)G(w,y) + O(\varepsilon^2).$$

Now we consider the following problem.

PROBLEM. Let Ω be a bounded domain in $\mathbb{R}^3$. Let $\widetilde{\Omega}_\varepsilon$ denote the domain $\Omega \setminus \bigcup_{i=1}^n B(\varepsilon; w_i)$. Let $G_\varepsilon(x,y)$ be the Green function associated with $\widetilde{\Omega}_\varepsilon$. Can one say that

$$G_\varepsilon(x,y) = G(x,y) - 4\pi\varepsilon \sum_{i=1}^n G(x,w_i)G(w_i,y) + O(\varepsilon^2)?$$

Unfortunately this is not true when $n \to \infty$ ($\varepsilon \to 0$). The term $s \geq 2$ in (5.1) must be added.

Next we explain (5.1) in terms of Brownian motion. It is well known that the Green function equals $\int_0^\infty e(t,x,y)\,dt$ by the heat kernel $e(t,x,y)$. When the Brownian path starts at x, it is killed at $\partial\Omega$. If we remove $B(\varepsilon,w_i)$, then a new killing boundary is added. This means that the paths that start from $\partial B(\varepsilon;w_i)$ to $\partial\Omega$ are negligible. If we remove these paths (N-path) we exclude too many paths.

We must add the paths that pass through $\partial B(\varepsilon; w_i)$ and $\partial B(\varepsilon; w_j)$ and that are killed at $\partial\Omega$. Then we have added too many paths. These correspond to

$$\sum_{i=1}^{n} G(x,w_i)G(w_i,y), \quad \sum_{\substack{i,j=1\\i\neq j}}^{n} G(x,w_i)G(w_i,w_j)G(w_j,y).$$

It is the so-called exclusion-inclusion formula for probability theory. However, we do not know the reason why we use a self-avoiding sum except through experience.

The reason why we use the terminology "point interaction approximation" is given by the following. We consider the Born expansion

$$(5.2) \quad G(x,y) + \varepsilon^* \int_\Omega G(x,z)U(z)G(z,y)\,dz$$
$$+ \sum_{k=2}(\varepsilon^*)^k \int \cdots \int_{\Omega^k} G(x,z_1)U(z_1)G(z_1,z_2)\cdots U(z_k,y)\,dz$$
$$(dz = dz_1\cdots dz_k).$$

We set $U = c\sum_{i=1}^n \delta(x-w_i)$. Then the formula (5.1) appears, and conversely, we can conclude the convergence of (5.1) to (5.2) with $\varepsilon^* = -4\pi\alpha m^{\beta-1}$, $U=V$ by the law of large numbers.

QUESTION AND ANSWER. There is a book by S. Albeverio, F. Gesztesz, R. Høgh-Krohn, and H. Holden, *Solvable models in quantum mechanics* (Springer). The authors consider the Schrödinger operator with point interaction potential $\delta(x-w_i)$. Is there any relation? No, our method is to approximate the hard sphere by the point interaction approximation potential. And everything is in this approximation.

What can one say when the removed subset is not a sphere? Let Ω be a bounded domain in $\mathbb{R}^3$. We fix $w \in \Omega$. Let D be a domain which contains w. Then we put $\varepsilon D = \{x; \varepsilon^{-1}(x-w) \in D\}$. Then we have the variational formula for Green's function of the Laplacian in [20]. We see that the term $\operatorname{cap}(D)$ appears. Thus, we can also treat the case of many obstacles. However, the author would like to restrict the discussion to the case of the sphere for the sake of simplicity.

6. Perturbation theory. III

We want to show how to obtain Proposition 5.1. Then we explain the method of Green's function.

PROPOSITION 6.1. *If $u \in C^\infty(\omega) \cap C^0(\overline{\omega})$ satisfies*

$$(-\Delta + \lambda)u(x) = 0, \quad x \in \omega,\ \lambda > 0,$$
$$u(x) = 0, \quad x \in \partial\Omega \cap \partial\omega,$$
$$|u(x)| \leq M_r, \quad x \in \partial B(\alpha/m; w_r),\ r=1,\ldots,n,$$

then

$$\|u\|_{L^2(\omega)} \leq (C/m)\lambda^{-1/4}\left\{\sum_{r=1}^n M_r^2 + \sum_{\substack{r,q=1\\r\neq q}}^n M_r M_q \exp(-\sqrt{\lambda/8}|w_r - w_q|)\right\}^{1/2}.$$

We can prove Proposition 6.1 by the Hopf maximum principle. We set $u = \mathbb{G}_{w(m)}f - \mathbb{H}_{w(m)}f$ for $f \in L^2(\omega)$. Then Proposition 6.1 is satisfied with $M_r = \max\{|H_{w(m)}f(x)|; x \in \partial B(\alpha/m; w_r)\}$. Therefore, we must estimate the terms $\sum M_r^2$, etc. We consider a probabilistic estimate. Thus, we take the expectation. It is easy to see that the calculus involves the multiproduct of the Green function. We set $G_I = G(w_{i_1}, w_{i_2}) \cdots G(w_{i_{s-1}}, w_{i_s})$ and $G_J = G(w_{j_1}, w_{j_2}) \cdots G(w_{i_{t-1}}, w_{j_t})$. Then $E(G_I G_J)$ or more complicated term appears. Notice that $i_1, \ldots, i_s$ and $j_1, \ldots, j_t$ are self-avoiding. However, $(i_1, \ldots, i_s)$ with $(j_1, \ldots, j_t)$ is not self-avoiding. If there are q-pairs of $i_k = j_{\sigma(k)}$, $k = 1, \ldots, q$, then we say that I and J have q-intersections, and we write $\#(I \cap J) = q$. We know $E(G_I G_J) = \sum_{q=0} \sum_{\#[I \cap J]=q} E(G_I G_J)$. By the calculation we know that the right-hand side can be estimated when $\beta \in [1, 3)$. To obtain Proposition 5.1, we need to estimate various kinds of multiple products of Green's function. We use the Hölder inequality to reduce various formulas to the fundamental type of the formula $E(G_I G_J)$. This takes about 20 pages in Ozawa [16].

Perturbation theory is complicated. However, perturbative calculus gave different kinds of results such as Theorem 2.2. Here the author offers to the probabilist the following question. Can you give a proof of Theorem 2.2 by probabilistic method?

The most important open problem is the determination of the fluctuation for $\beta \in [1, 3)$. Now we can write

$$\mu_j(w(m)) = \mu_j + 4\pi\alpha m^{\beta-1}|\Omega|^{-1} + m^{(\beta/2)-1}\Pi_j.$$

We see that this formula does not hold for $\beta > 2$. If $E(\Pi_j^2) > E(\Pi_{j+1}^2)$, then this contradicts the fact that $\mu_j(w(m)) \leq \mu_{j+1}(w(m))$. The author conjectures that Theorem 4.1 is valid for $\beta < 2$, and for $\beta \geq 2$ the result is quite different. Its determination is an interesting problem in mathematical physics.

When $\dim \Omega \geq 4$, perturbative calculus is very complicated. The multiproduct of Green's function diverges when $d \geq 4$, if we calculate $E(G(w_i, w_j)^2)$. However, we can manage the integration by removing a neighborhood of the diagonal, and we can get the following result.

We remove the ball of radius $1/m$ with the centers $w_1, \ldots, w_n$, $n = [m^\beta]$ from the bounded domain Ω in $\mathbb{R}^4$. $\Omega_{w(m)}, \mu_j(w(m))$ are as stated above. Let $\mu_j(V; m)$ be the jth eigenvalue of the Schrödinger operator $-\Delta + 2\pi^2 m^{\beta-2}V$ in Ω under the Dirichlet condition on $\partial \Omega$. Then we have the following

THEOREM 6.2. *Fix $\beta \in [2, 12/5)$. Then*

$$\lim_{m \to \infty} P(w(m) \in \Omega^n; m^{(6-(5/2)\beta-\varepsilon)}|\mu_j(w(m)) - \mu_j(V; m)| < m^{\beta-2}) = 1$$

holds for $\varepsilon > 0$.

If $\beta \in [2, 12/5)$, then $6 - (5/2)\beta > 0$. If $\beta = 2$, then

$$|\mu_j(w(m)) - \mu_j(V; m)| < m^{-1+\varepsilon}.$$

We believe this may be the best estimate, since the fluctuation may be of type $m^{-1}\Phi_j$. However, we do not know whether the fluctuation is Gaussian or not. For Theorem 6.2, the reader is referred to [19].

For $n \geq 5$, we have no perturbative result at the present time.

Now we consider the general case heuristically. Let Ω be a bounded region in $\mathbb{R}^d$ ($d \geq 3$). Let $\Omega_\varepsilon = \Omega \backslash \varepsilon$-ball, let the ε-ball's center be w. Let $\mu_j(\varepsilon)$ be the jth eigenvalue of $-\Delta$ in Ω_ε under the Dirichlet condition. This is written as

$$\mu_j(\varepsilon) - \mu_j = c\varepsilon^{d-2}\varphi_j(w)^2 + c'\varepsilon^{d-1}g_j(w) + \cdots.$$

We remove m^{d-2} balls of radius $1/m$ with centers $w_1, \ldots, w_n$ ($n = m^{d-2}$) from Ω. We can also have

$$\mu_j(w(m)) - (\mu_j + C|\Omega|^{-1}) \to 0$$

as the Lenz shift. In that case if the fluctuation is Gaussian, then it is of order $m^{-(d-2)/2}$. However, $c'm^{-(d-1)}m^{d-2}$ is greater than $m^{-(d-2)/2}$ for $d \geq 5$. Thus, the fluctuation largely depends on $\dim \Omega$ and on β.

Next we comment on another important problem. For the sake of simplicity let $\dim \Omega = 3$. Let $\varphi_1(w(m))$ be the L^2-normalized eigenfunction (of course, in ω). Does there exist $C > 0$ such that

$$\lim_{m \to \infty} P\left(\sup_{x \in \omega} |\varphi_1(w(m))| < C\right) = 1?$$

THEOREM 6.5 ([12]). *Let u be the first eigenfunction of the Laplacian in $\Omega \subset \mathbb{R}^3$. Then*

$$|u(x)| \leq (2\pi^{1/2})^{-1} \min\{\lambda(R/3)^{1/2}, \lambda^{3/2} R^{3/2}/(45)^{1/2}\}.$$

Here λ is the first eigenvalue, $R = ((3/4)|\Omega|)^{1/3}$.
Then $|\omega| \in (|\Omega|/2, |\Omega|)$ implies $|\varphi_1(w(m))| \leq C'$.

We state the following conjecture.

CONJECTURE. *Fix j. Fix $\beta \in [1, 3)$. Then there exists a constant C such that*

$$\lim_{m \to \infty} P(|\varphi_j(w(m))| < C) = 1.$$

This conjecture is related to (4.2), and it will be an important problem.

QUESTION AND ANSWER. What can one say about $\beta = 3$? Unfortunately, perturbative calculus cannot be applied. The theory of the Wiener sausage again fits the case $\beta = 3$. If $\dim = n$, a representation of the limit

$$\lim_{t \to \infty} t^{-n/(n+2)} \log E(\exp(-|W_\delta(t, \omega)|))$$

is given by Donsker-Varadhan [33]. By this theory we can say a little about eigenvalues. We think that the case $\beta \nearrow 3$ is the most difficult case of our investigation.

We offer the following problem to the specialist in probability theory.

PROBLEM. Determine the fluctuation of

$$\sum_{j=1}^{\infty} e^{-\mu_j(w(m))t}.$$

We mention an interesting paper by Chavel-Feldman [2] that uses the Wiener sausage to obtain Kac's result.

7. A nonlinear problem

Next we discuss the problem of singular variation of domains. Let $\Omega \subset \mathbb{R}^2$ be a bounded domain with smooth boundary $\partial\Omega$, $w \in \Omega$. We set $\Omega_\varepsilon = \Omega \setminus \overline{B(\varepsilon;w)}$. We write Ω as Ω_0. We consider the following minimizing problem

$$(7.1)_\varepsilon \qquad \mu(\varepsilon) = \inf_X \left(\int_{\Omega_\varepsilon} |\nabla u|^2 \, dx + k \int_{\partial B_\varepsilon} u^2 \, d\sigma_x \right),$$

where $B_\varepsilon = B(\varepsilon;w)$, $X = \{u \in H^1(\Omega_\varepsilon), u = 0 \text{ on } \partial\Omega, \|u\|_{L^{p+1}(\Omega_\varepsilon)} = 1\}$. We determine $\mu(0)$ by $\Omega_0 = \Omega$, $\partial B_0 = \varnothing$. It is easy to show that, for $p \in (1,\infty)$, there exists at least one function $u_\varepsilon \in X$ which attains (7.1) and satisfies

$$-\Delta u_\varepsilon(x) = \mu(\varepsilon) |u_\varepsilon|^{p-1} u_\varepsilon(x), \qquad x \in \Omega_\varepsilon,$$

$$u_\varepsilon(x) = 0, \qquad x \in \partial\Omega,$$

$$k u_\varepsilon(x) + \frac{\partial}{\partial \nu_x} u_\varepsilon(x) = 0, \qquad x \in \partial B_\varepsilon.$$

We have the following.

THEOREM 7.1 ([14]). *Assume that the solution that attains* $(7.1)_\varepsilon$ *is unique up to its signature for* $0 \leq \varepsilon \ll 1$. *Assume that* $\mathrm{Ker}(\Delta + p\mu(\varepsilon)|u_\varepsilon|^{p-1}) = 0$ *for* $0 < \varepsilon \ll 1$, *plus some conditions. Then* $\mu(\varepsilon) - \mu(0) = 2\pi k \varepsilon u(w)^2 + o(\varepsilon)$.

See [14] for some conditions.

This result says that the singular variational formula holds for nonlinear eigenvalues. What can one say about eigenvalues when the number of disks which are removed tends to ∞ and the radius ε tends to 0 in the limit?

8. The heat equation

What can one say about heat equation when the domain has many holes? Let us follow Papanicolaou-Varadhan [22]. For the sake of simplicity we treat the case where the obstacles are spheres. Let $m = 1, 2, \ldots$ be a parameter. We remove m balls of radius $1/m$ with the centers $w_1, \ldots, w_m$ from $\mathbb{R}^3$, and we write $\Omega_{w(m)} = \mathbb{R}^3 \setminus m\text{-balls}$. We consider the following equation

$$\frac{\partial}{\partial t} u^{(m)}(x,t) = (1/2) \Delta u^{(m)}(x,t), \qquad t > 0,\ x \in \Omega_{w(m)},$$

(8.1) $\qquad u^{(m)}(x,t) = 0, \qquad t > 0,\ x \in \partial\Omega_{w(m)},$

$$u^{(m)}(x,0) = f(x), \qquad x \in \Omega_{w(m)}.$$

Now we make the following assumptions. Let $V(x) \geq 0$ be a continuous function with compact support in $\mathbb{R}^3$. We assume that

$$\lim_{m \to \infty} m^{-1} \sum_{i=1}^m \varphi(w_i) = \int_{\mathbb{R}^3} \varphi(x) V(x) \, dx,$$

$$\lim_{m \to \infty} m^{-2} \sum_{\substack{i,j=1 \\ i \neq j}}^m |w_i - w_j|^{-1} = \int_{\mathbb{R}^3} \int_{\mathbb{R}^3} V(x) V(y) |x-y|^{-1} \, dx \, dy.$$

Then we have the following.

THEOREM 8.1 ([22]). *Under the above assumptions we have the following. Let $\varepsilon > 0$, $T > 0$ be arbitrary numbers. Then there exists m_0 such that, for any $m \geq m_0$, there exists a subset $\Omega_\varepsilon^{(m)}$ of $\Omega_{w(m)}$ satisfying $\mathrm{vol}(\mathbb{R}^3 \setminus \Omega_\varepsilon^{(m)}) < \varepsilon$ and*

$$\sup_{0 \leq t \leq T} \sup_{x \in \Omega_\varepsilon^{(m)}} |u^{(m)}(x,t) - u(x,t)| < \varepsilon.$$

Here $u(x,t)$ is a solution of

$$u_t = (1/2)\Delta u - 2\pi V u, \quad t > 0,$$
$$u(x,0) = f(x), \quad x \in \mathbb{R}^3.$$

This result is a deterministic version of the result. The probabilistic version is Theorem 8.2. Let $w_1, \ldots, w_m$ be independent identically distributed random variables. We fix $0 \leq V \in C_0^0(\mathbb{R}^3)$ such that $\int_{\mathbb{R}^3} V(x)\, dx = 1$. Then, $\mathbb{R}^3$ can be considered as a probability space.

THEOREM 8.2 ([22]). *For any $\varepsilon > 0$, $\delta > 0$, $T < \infty$, there exists m_0 such that for any $m \geq m_0$ there exists $\Omega_{\varepsilon,\delta}^{(m)} \subset (\mathbb{R}^3)^m$, $\Omega_\varepsilon^{(m)} \subset \Omega_{w(m)}$, and $\omega \in \Omega_{\varepsilon,\delta}^{(m)}$ satisfying*

$$\sup_{0 \leq t \leq T} \sup_{x \in \Omega_\varepsilon^{(m)}} |u^{(m)}(x,t) - u(x,t)| < \varepsilon, \qquad \omega \in \Omega_{\varepsilon,\delta}^{(m)},$$

$$\mathrm{vol}(\mathbb{R}^3 \setminus \Omega_\varepsilon^{(m)}) < \varepsilon,$$

and

$$\mathrm{Prob}((\mathbb{R}^3)^m \setminus \Omega_{\varepsilon,\delta}^{(m)}) < \delta.$$

Here $u(x,t)$ is the same as in Theorem 8.1.

For related topics see [1]. This is a natural problem when the hole is not a killing boundary (third condition).

9. Smoluchowski theory

We follow Lang-Nguyen X. X. [11], and we introduce the theory of coagulation (by Smoluchowski). In $\mathbb{R}^3$ there are m-balls of radius $R/2$. The particles are moving independently as a Brownian particle of diffusion constant $1/2$. If the distance of the centers of two particles becomes R, a particle i disappears with probability $1/2$ and a particle k disappears with probability $1/2$. Then one of the particles continues in a Brownian motion of diffusion constant $1/2$. Let $m \to \infty$, $R \to 0$ while $mR = 1$ always.

Let $u(x,t)$ be the density of particles. Assume that

$$m^{-1} \sum_{i=1}^m \varphi(x_i) \to \int_{\mathbb{R}^3} \varphi(x) u(x,0)\, dx.$$

The problem is to determine $u(x,t)$. Here we want to determine $u(x,t)$ heuristically. The solution is

(9.1)
$$\frac{\partial}{\partial t} u(x,t) - (1/2)\Delta u(x,t) + 2\pi u(x,t)^2 = 0,$$
$$\lim_{t \to 0} u(x,t) = u(x,0).$$

If there are m particles around a fixed labeled particle, then this particle appears to move in a field of $-(1/2)\Delta u + 2\pi u\cdot$. However, u itself is the limit density of the existence of a particle. Thus, (9.1) holds. This is mathematically studied by Lang-Nguyen X. X. [11], and Sznitman [26], [27], [28].

What can we say when the particle is reflecting or is the third kind particle? Saisho-Tanaka [24] has treated finite reflecting particles.

10. Multiple wave scattering

This is a mathematically interesting topic, although there is no rigorous theory as far as the author knows. We follow the paper by J. B. Keller, *Stochastic equation and wave propagation in random media*, Proceedings of the Sixteenth Symposium in Applied Mathematics, Amer. Math. Soc., Providence, R.I., 1964.

There are scatters in a medium. The wave comes and goes away. The number of scatters tends to infinity and its radius tends to zero in some scaling. What can one say about the wave equation $(\Delta + k^2)u(x) = 0$, $k > 0$? If the center of the scatters is considered as a random variable, then it is a problem of random media. A formalism to discuss the random media with wave equation is discussed in J. B. Keller. However, this formalism is not rigorous.

The fundamental paper in this direction is M. Lax: Rev. Modern Phys. **23** (1951). The author hopes that these results can be treated in a mathematically rigorous way.

References

1. J. R. Baxter, R. V. Chacon, and N. C. Jain, *Weak limits of stopped diffusion*, Trans. Amer. Math. Soc. **293** (1986), 767–792.
2. I. Chavel and E. A. Feldman, *The Lenz shift and Wiener sausage in insulated domains*, From Local Time to Global Geometry, Control and Physics (K. D. Elworthy, ed.), Longman, Harlow, Scientific and Technical, 1984/85, pp. 47–66.
3. D. Cioranescu and F. Murat, *Un term étrange venu d'aillerus*, Nonlinear Partial Differential Equation, Res. Notes Math., vol. 60, Boston, Mass., 1982, pp. 98–138.
4. G. Dal Masso, E. De Giorgi, and L. Modica, *Weak convergence of measures on spaces of lower semicontinuous functions*, Suppl. Rend. Circ. Math. Palermo Serie (2) **15** (1987), 59–100.
5. R. Figari, E. Orlandi, and S. Teta, *The Laplacian in region with many obstacles, fluctuation around the limit operator*, J. Statist. Phys. **41** (1985), 465–487.
6. J. Fröhlich and T. Spencer, *Absence of diffusion in the Anderson tight binding model for large disorder or low energy*, Comm. Math. Phys. **88** (1983), 151–184.
7. E. Ja. Huruslov and V. A. Marchenko, *Boundary value problem in region with fine grained boundaries*, Kiev, 1974. (Russian)
8. M. Kac, *Probabilistic methods in some problem of scattering theory*, Rocky Mountain J. Math. **4** (1974), 511–537.
9. S. Kotani, *Ljapunov indices determine absolutely continuous spectra of random one-dimensional Schrödinger operator*, Proc. Taniguchi Sympos., Katata 1982, pp. 225–247.
10. S. Kaizu, *The Poisson equation with semilinear boundary conditions in domains with many tiny holes*, J. Fac. Sci. Univ. Tokyo Sect. IA Math. **36** (1989), 43–86.
11. R. Lang and Nguyen X. X., *Smoluchowski's theory of coagulation in the colloids holds rigorously in the Boltzmann-Grad-limit*, Random Fields Vol. II, Colloq. Math. Soc. Janos Bolyai (J. Fritz, J. L. Lebowitz, and D. Szasz, eds.), vol. 27, North-Holland, Amsterdam, 1981.
12. E. Mascolo, L. Migliaccio, and R. Schianchi, *An inequality for L^∞-norm of eigenfunctions of linear second order elliptic operators*, Boll. Anal. Funzionale Appl. **1** (1982), 51–60.
13. H. Osada, *Homogenization of reflecting barrier Brownian motions*, Proc. Taniguchi workshop at Sanda and Kyoto, 1990, Pitman Res. Notes Math. Ser., Longman Scientific and Technical, Harlow (to appear).
14. T. Osawa and S. Ozawa, *Nonlinear eigenvalue problems and singular variation of domains*, Kodai Math. J. **15** (1992), 313–323.

15. S. Ozawa, *Point interaction potential approximation for $(-\Delta + U)^{-1}$ and eigenvalues of the Laplacian on wildly perturbed domain*, Osaka J. Math. **20** (1983), 923–937.
16. _____, *Fluctuation of spectra in random media*. II, Osaka J. Math. **27** (1990), 17–66.
17. _____, *Singular variation of domain and spectra of the Laplacian with small Robin conditional boundary*. I, Osaka J. Math. **29** (1992), 405–418.
18. _____, *Random media with many randomly distributed obstacles with the third boundary conditions*, preprint.
19. _____, *Spectra of random media with many randomly distributed obstacles*, Osaka J. Math. **30** (1993), 1–27.
20. _____, *Electrostatic capacity and eigenvalues of the Laplacian*, J. Fac. Sci. Univ. Tokyo Sect. IA Math. **30** (1983), 53–62.
21. S. Ozawa and S. Roppongi, *Singular variation of domain and spectra of the Laplacian with small Robin conditional boundary*. II, Kodai Math. J. **15** (1992), 403–429.
22. G. C. Papanicolau and S. R. S. Varadhan, *Diffusion in region with many small holes*, Lecture Notes in Control and Inform. Sci., vol. 75, Springer-Verlag, Berlin and New York, 1980, pp. 190–206.
23. J. Rauch and M. Taylor, *Potential and scattering theory on wildly perturbed domain*, J. Funct. Anal. **18** (1975), 27–59.
24. Y. Saisho and H. Tanaka, *Stochastic differential equations for mutually reflecting Brownian balls*, Osaka J. Math. **23** (1986), 725–740.
25. B. Simon, *Functional integration and quantum physics*, Academic Press, New York, 1979.
26. A. S. Sznitman, *Some bounds and limiting results for the measure of Wiener sausage of small radius associated to elliptic diffusions*, Stochastic Process. Appl. **25** (1987), 1–25.
27. _____, *A limiting result for the structure of collision between many independent diffusion*, Probab. Theory Related Fields **81** (1989), 353–381.
28. _____, *Propagation of chaos for a system of annihilating Brownian spheres*, Comm. Pure Appl. Math. **60** (1987), 663–690.
29. Y. Takahashi, *An integral representation on the path space for scattering length theory*, Osaka J. Math. **27** (1990), 373–379.
30. M. Taylor, *Scattering length and perturbation of $-\Delta$ by positive potentials*, J. Math. Anal. Appl. **53** (1976), 291–312.
31. J. M. Zaiman, *Models of disorder*, Cambridge Univ. Press, London and New York, 1979.
32. V. V. Zikov, S. M. Kozlov, O. A. Oleinik, and Kha Ten Nagon, *Averaging and G-convergence of differential operators*, Russian Math. Surveys **34** (1979), 69–147.
33. M. Donsker and S. R. S. Varadhan, *Asymptotic for the Wiener sausage*, Comm. Pure Appl. Math. **28** (1975), 525–565.
34. M. M. Schiffer and D. C. Spencer, *Functional of finite Riemann surfaces*, Princeton Univ Press, Princeton, NJ, 1954.
35. S. Kaizu, *The Poisson equation with non-autonomous semi-linear boundary condition in domains with many tiny holes*, SIAM J. Math. Anal. **22** (1991), 1222–1245.

Translated by SHIN OZAWA

The Asymptotic Distributions of Eigenvalues for the Schrödinger Operators with Magnetic Fields

Hiroyuki Matsumoto

It might be folklore that a starting point for the study on the asymptotic distributions of eigenvalues for differential operators is the lecture given by the physicist A. Lorentz in 1910 at Göttingen. Even if we restrict ourselves to the study of Schrödinger operators $-\Delta/2 + V$, many authors have been studying this problem since the paper by de Wet and Mandl [7] and the Titchmarsh's works [42, 43], which are the oldest works known to the author. Also for the Schrödinger operators with magnetic fields, it has recently been found that, if the magnitude of the magnetic fields grows unboundedly at infinity, they have compact resolvents and their spectra consist only of the eigenvalues with finite multiplicities, and the asymptotic distributions of the eigenvalues have been studied for such Schrödinger operators.

In this article, keeping in mind the idea by M. Kac of using the path integrals on the Wiener space, I would like to discuss the asymptotic distributions of the eigenvalues mainly for the Schrödinger operators with magnetic fields.

1. Introduction

R. P. Feynman reformulated the nonrelativistic quantum mechanics by developing the theory of path integrals. His proposal begins with the path integral representation of the solutions of the initial value problems for the Schrödinger equations and covers the problems in quantum mechanics quite extensively. The theory and the applications of path integral are mentioned in detail in the book by Feynman himself [10]. In this Sûgaku, Fujiwara–Asada [11] and Ichinose [14] have written on the mathematical approach to the path integral. Moreover it is known that Feynman's idea has been influencing many fields of mathematics, for example, analysis on the loop spaces, the Malliavin calculus and so on.

On the other hand, Kac found the usefulness of the arguments on Wiener space (a path space) similar to the formal deduction in the theory of the path integral. He used this idea to study the problem of a drum or Weyl's problem, i.e., the asymptotic distribution of the eigenvalues for the Laplacian on bounded domains [24].

Kac found furthermore that the semigroup whose infinitesimal generator is a Schrödinger operator can be represented by means of the integration (expectation) of some Wiener functional. This representation is called the Feynman–Kac formula

---

1991 *Mathematics Subject Classification*. Primary 35J10.
This article originally appeared in Japanese in Sûgaku **44** (4) (1992), 320–329.

and, by replacing the Schrödinger equations with the heat equations, it realizes with mathematical rigor Feynman's idea of representing the solutions in terms of the integrations on a path space. This formula is indispensable when we apply the integration of Wiener functionals not only to the analysis of the Schrödinger operators but also to problems in differential geometry and so on.

In this article we consider the Schrödinger operators which have potentials growing unboundedly at infinity and whose spectra consist only of the eigenvalues with finite multiplicities. For such Schrödinger operators D. B. Ray [33] has studied the asymptotic distributions of the eigenvalues by using Kac's idea. Although Kac himself has referred to Ray's result in his books [22, 24], I give the main point here in order to make the arguments in the subsequent sections easier to understand.

Letting V be a real-valued function on the d-dimensional Euclidean space $\mathbf{R}^d$ satisfying

$$(1.1) \qquad \lim_{|x| \to \infty} V(x) = \infty$$

and assuming suitable continuity of V, we consider the essentially selfadjoint Schrödinger operator H_1 defined by $H_1 = -\frac{1}{2}\Delta + V$, $\Delta = \sum_{i=1}^d (\partial/\partial x^i)^2$. We denote by the same notation its selfadjoint extension to $L^2(\mathbf{R}^d)$, which is a selfadjoint operator bounded from below. Its spectrum consists only of eigenvalues with finite multiplicities [34]. We denote the eigenvalues by $\{\lambda_n\}_{n=1}^\infty$ counting the multiplicities and set $N_1(\lambda) = \#\{n; \lambda_n < \lambda\}$. The purpose is to study the asymptotic behavior of $N_1(\lambda)$ as $\lambda \uparrow \infty$.

We define the path space $W_0(\mathbf{R}^d)$ by

$$W_0(\mathbf{R}^d) = \{\omega : [0, \infty) \ni t \mapsto \omega(t) \in \mathbf{R}^d; \omega \text{ is continuous and } \omega(0) = 0\}$$

and, by endowing $W_0(\mathbf{R}^d)$ with the Wiener measure, we define the d-dimensional Wiener space. We denote by $p(t, x, y)$ the integral kernel of the semigroup $\exp(-tH_1)$, $t > 0$, generated by H_1. $p(t, x, y)$ is the heat kernel for H_1. Then the Feynman–Kac formula says that the solution u for the heat equation

$$\frac{\partial}{\partial t} u = -H_1 u, \qquad u|_{t=0} = f,$$

can be written in the following form:

$$u(t, x) = \int_{\mathbf{R}^d} p(t, x, y) f(y) \, dy = E_0 \left[f(x + \omega_t) \exp\left\{ -\int_0^t V(x + \omega_s) \, ds \right\} \right],$$

where E_0 is the expectation (integral) with respect to the Wiener measure. As is seen by replacing f with δ_x formally (this can be justified by the help of the Malliavin calculus, see [44]), the heat kernel $p(t, x, y)$ can be expressed as

$$p(t, x, x) = (2\pi t)^{-d/2} E_0 \left[\exp\left\{ -\int_0^t V(x + \omega_s) \, ds \right\} \,\middle|\, \omega_t = 0 \right]$$

on the diagonal set, where $E_0[\,\cdot\,|A]$ denotes the conditional expectation with respect to the Wiener measure under the condition A. Moreover, by the time homogeneity of the Wiener measure (the probability laws of $\{\omega_s\}_{s\geq 0}$ and $\{\varepsilon^{-1/2}\omega_{\varepsilon s}\}_{s\geq 0}$ are identical for every $\varepsilon > 0$), we have

$$(1.2) \quad p(t,x,x) = (2\pi t)^{-d/2} E_0 \left[\exp\left\{ -t \int_0^1 V(x + \sqrt{t}\omega_s) ds \right\} \bigg| \omega_1 = 0 \right].$$

From this expression and the continuity of V, it is easy to see

$$(1.3) \quad p(t,x,x) = (2\pi t)^{-d/2} e^{-tV(x)} \cdot (1 + o(1)) \qquad \text{as } t \downarrow 0.$$

Now the counting function $N_1(\lambda)$, which is the object of our study, is a nondecreasing function, and the positive measure $dN_1(\lambda)$ on $\mathbf{R}$ is defined in a natural manner. The Laplace transform of $dN_1(\lambda)$ is equal to the trace of the semigroup $\exp(-tH_1)$, and it holds that

$$\int_{\mathbf{R}} e^{-t\lambda} dN_1(\lambda) = \mathrm{Tr}(e^{-tH_1}) = \int_{\mathbf{R}^d} p(t,x,x)\, dx,$$

where dx is the d-dimensional Lebesgue measure. Therefore, if we assume that $\exp(-tV)$ is integrable on $\mathbf{R}^d$ for every $t > 0$ and set

$$(1.4) \quad \begin{aligned} I(t) &= (2\pi t)^{-d/2} \int_{\mathbf{R}^d} e^{-tV(x)} dx \\ &= (2\pi)^{-d} \int_{\mathbf{R}^d \times \mathbf{R}^d} \exp\left\{ -t\left(\frac{1}{2}|p|^2 + V(x)\right)\right\} dx\, dp, \end{aligned}$$

it is natural to expect that

$$(1.5) \quad \mathrm{Tr}(e^{-tH_1}) = I(t) \cdot (1 + o(1))$$

holds as $t \downarrow 0$ by (1.3). But, since we cannot expect the uniformness in x of the small order $o(1)$ in the right-hand side of (1.3), some rigorous estimates are necessary in order to prove (1.5). Ray has proved (1.5) by using the Feynman–Kac formula and the properties of the Wiener processes under some suitable assumptions on the continuity of V. Once (1.5) is proved, the Tauberian theorem implies that the following equality holds as $\lambda \uparrow \infty$:

$$(1.6) \quad N_1(\lambda) = (2\pi)^{-d} \mathrm{vol}(\{(x,p)\in \mathbf{R}^d \times \mathbf{R}^d; \tfrac{1}{2}|p|^2 + V(x) < \lambda\}) \cdot (1 + o(1)).$$

Although the proof of (1.5) was not mentioned, the above is the outline of Ray's result, and we can see part of Kac's idea there. Also for the Schrödinger operators with magnetic fields, the inference along this line is an important step for the study of the asymptotic distributions of the eigenvalues.

In this article we consider the Schrödinger operator $H = H_{(\theta,V)}$ on $\mathbf{R}^d$ that is defined by

$$(1.7) \quad H = H_\theta + V, \quad H_\theta = \frac{1}{2} \sum_{j=1}^d \left(\sqrt{-1}\frac{\partial}{\partial x^j} - \theta_j\right)^2.$$

Following the usual terminology, we call $\theta = (\theta_1, \theta_2, \ldots, \theta_d): \mathbf{R}^d \to \mathbf{R}^d$ a vector potential and $V: \mathbf{R}^d \to \mathbf{R}$ a scalar potential. The vector potential is identified with the differential form of the first degree defined by $\theta = \sum_{j=1}^d \theta_j dx^j$. Differential forms of nth degree will be simply called n-forms below. Then the 2-form $d\theta$ or

$\mathrm{curl}(\theta)(x) = (\partial_j \theta_i(x) - \partial_i \theta_j(x))_{i,j=1,2,\ldots,d}$, a function valued in $d \times d$ real skew-symmetric matrices, represents the corresponding magnetic field and the spectrum of H is determined by $d\theta$ and V not by θ by virtue of the gauge invariance.

We will concentrate our attention on the asymptotic distribution of the eigenvalues of H. About the various topics on the Schrödinger operators with magnetic fields, for example, the gauge invariance, the supersymmetry and so on, we refer the readers to B. Simon's survey in his lecture note [6], where the detailed references are also given.

Generally speaking, if we consider the complex line bundle over a Riemannian manifold, the connection form which gives us the connection on the bundle determines the covariant Laplacian. We can define the operator H_θ as a covariant Laplacian, which has a geometrical meaning. Under certain assumptions on the base manifold, we can prove the main theorem (Theorem 2.1) of this article in this framework [17, 29]. But, in order to make the arguments clear, we consider the Schrödinger operator defined on the Euclidean space $\mathbf{R}^d$ endowed with the usual metric.

In order to discuss the asymptotic distribution of the eigenvalues, we have to know, at first, when the spectrum of H consists only of the eigenvalues with finite multiplicities and under which assumptions on the potentials the resolvents of H are compact operators. The problem occurs when (1.1) fails. There arise some delicate problems, and they themselves are interesting. Roughly speaking, if $V(x) + \|d\theta\|(x)$ diverges at infinity, the resolvents of H are compact. But the problem is not so simple. For details, see the next section, [3, 8, 12, 13, 22] and so on.

In the following, I would like to mention the known results on the asymptotic distribution of the eigenvalues for H in the framework described above. We will follow mainly [17, 29] and will refer to some related topics.

2. The trace of the semigroup

Let H be the Schrödinger operator defined by (1.7). In this section, following [17, 29], we will mention the asymptotic behavior of the trace $\mathrm{Tr}(\exp(-tH))$ of the semigroup $\exp(-tH)$ as $t \downarrow 0$.

To make the arguments easier, we assume the following:
(H.1) V, θ_j, $j = 1, 2, \ldots, d$, are C^∞ real-valued functions on $\mathbf{R}^d$ and V is bounded from below.

Next, setting

$$m(\lambda) = \mathrm{vol}(\{x \in \mathbf{R}^d; h(x) < \lambda\}), \quad h(x) = V(x) + \tfrac{1}{2}\|d\theta\|(x),$$

we assume the following:
(H.2) $\lim_{|x|\to\infty} h(x) = \infty$;
(H.3) There exists a positive constant C such that $m(2\lambda) \leq Cm(\lambda)$ holds for sufficiently large λ;
(H.4) For every $i, j, k, l = 1, 2, \ldots, d$, it holds that

$$|\partial_i V(x)| + |\partial_k \partial_l \theta_j(x)| = o(h(x)^{3/2})$$

as $|x| \to \infty$, where $\partial_i = \partial/\partial x^i$.

Under the assumptions above, H is essentially selfadjoint on $C_0^\infty(\mathbf{R}^d)$. We will denote by the same notations the selfadjoint extensions on $L^2(\mathbf{R}^d)$ of essentially selfadjoint operators.

I would like to remark on the assumptions (H.3) and (H.4). Roughly speaking, (H.3) says that the growth of h is faster than a certain polynomial in $|x|$. (H.4) might look like only a technical assumption at first glance but it is not necessarily right. When $d \geq 3$ and (1.1) fails, the resolvents of H are not compact operators in general if we replace the right hand side by $o(h(x)^\delta)$, $\delta > 2$ (Iwatsuka [22]). This might not be a big problem but is an interesting one. The author has an idea that the large oscillation of the magnitude of the magnetic field yields a set where the magnetic field is not very strong and that, through such a set, the particle can reach far away. It is a matter of course that (H.4) plays an important role when we estimate the error term. Under the assumption (H.4), the asymptotic distributions of the eigenvalues have been studied by Titchmarsh [43] when $\theta = 0$, $d = 1$ and by Colin de Verdière [5], Tamura [41] when $V = 0$.

At any rate, under the assumptions (H.1)–(H.4), we can prove that the semigroup $\exp(-tH)$ is of trace class for every $t > 0$ [29, 41] and we can study the asymptotic behavior of its trace as $t \downarrow 0$.

In order to state the result, we introduce some notation. We let $2 \cdot r(x)$ be the rank of the real skew-symmetric matrix $\mathrm{curl}(\theta)(x)$ and denote its eigenvalues by $\pm\sqrt{-1}b_k(x)$, $k = 1, \ldots, [d/2]$. When d is odd, 0 is always an eigenvalue and we include it. This means that, choosing a suitable orthonormal basis $\langle e^1, e^2, \ldots, e^d \rangle$ of the cotangent space $T_x^*(\mathbf{R}^d)$ at x, $d\theta(x) \in T_x^*(\mathbf{R}^d) \wedge T_x^*(\mathbf{R}^d)$ can be written as

$$d\theta(x) = \sum_{j=1}^{r(x)} b_j(x) e^{2j-1} \wedge e^{2j}.$$

Moreover, by (H.1), we can assume without loss of generality that

$$b_1(x) \geq \cdots \geq b_{r(x)}(x) > 0 = b_{r(x)+1}(x) = \cdots = b_{[d/2]}(x)$$

and that $b_k(x)$'s are continuous in x.

Then the following holds:

THEOREM 2.1. *If V and θ satisfy* (H.1)–(H.4), *the semigroup $\exp(-tH)$ is of trace class for every $t > 0$, and it holds that*

$$(2.1) \qquad \mathrm{Tr}(e^{-tH}) = (2\pi t)^{-d/2} \int_{\mathbf{R}^d} e^{-tV(x)} \prod_{j=1}^{[d/2]} \frac{tb_j(x)/2}{\sinh tb_j(x)/2} \, dx \cdot (1 + o(1))$$

as $t \downarrow 0$, where $a/\sinh a$ is replaced by 1 if $a = 0$. In particular, if $d = 2$ or 3 and if the norm of $d\theta$ is denoted by $\|d\theta\|$, then it holds that

$$\mathrm{Tr}(e^{-tH}) = (2\pi t)^{-d/2} \int_{\mathbf{R}^d} e^{-tV(x)} \frac{t\|d\theta\|(x)/2}{\sinh t\|d\theta\|(x)/2} \, dx \cdot (1 + o(1)).$$

3. Asymptotic distribution of eigenvalues

In this section, using Theorem 2.1, I would like to discuss the asymptotic distribution of the eigenvalues for the Schrödinger operator H defined by (1.7). Unfortunately, I have to remark first that we have not succeeded in expressing the asymptotic distribution concretely by using the Tauberian theorem and (2.1) in general. But, by virtue of the results of Ray, Colin de Verdière and Tamura, we have the concrete expressions for the asymptotic distributions of the eigenvalues like (1.6) and (3.2) below when $\theta = 0$ or $V = 0$. Moreover we can show by (2.1) that the problem is reduced to one of these cases if the growth rates of the two potentials are different. I will mainly deal with such situations. In the final part of this section, some examples will be given.

In order to mention the first result, we set $H_1 = -\frac{1}{2}\Delta + V$ and let I be the function on $(0, \infty)$ defined by (1.4). Furthermore, we denote by $\{\lambda_n\}_{n=1}^{\infty}$ the eigenvalues of H counting the multiplicities and set $N(\lambda) = \#\{n; \lambda_n < \lambda\}$.

Then we can show the following by using (2.1) and the Tauberian theorem.

COROLLARY 3.1. *Under the assumptions* (H.1)–(H.4), *if* $\|d\theta(x)\| = o(V(x))$ *as* $|x| \to \infty$, *it holds that*

$$\mathrm{Tr}(e^{-tH}) = \mathrm{Tr}(e^{-tH_1}) \cdot (1 + o(1)) = I(t) \cdot (1 + o(1))$$

as $t \downarrow 0$. *Moreover, if* $I(t)$ *is regularly varying as* $t \downarrow 0$, (1.6) *or*

(3.1) $\quad N(\lambda) = (2\pi)^{-d}\mathrm{vol}(\{(x, p); \frac{1}{2}|p - \theta(x)|^2 + V(x) < \lambda\}) \cdot (1 + o(1))$

holds as $\lambda \uparrow \infty$.

When the magnetic field is uniform, a similar assertion to that of this corollary has been proved by Odencrantz [31] by using the theory of pseudodifferential operators. I would like to mention here that, in Corollary 3.1, we have treated the case where we can ignore the stochastic oscillatory integral, which will be mentioned in §4. In [27] the author has shown the same assertion together with the calculation of the second term in some special cases using only rough estimates which are obtained by elementary probabilistic arguments.

Next let us consider the case where the scalar potential does not affect the leading term of the asymptotic distribution. We let N_0 be the right continuous function on $(0, \infty)$ defined as follows:

$$\int_0^{\infty} e^{-t\lambda}\,dN_0(\lambda) = J(t) \equiv (2\pi t)^{-d/2} \int_{\mathbf{R}^d} \prod_{j=1}^{[d/2]} \frac{tb_j(x)/2}{\sinh tb_j(x)/2}\,dx, \quad t > 0.$$

If the rank of $\mathrm{curl}(\theta)(x)$ does not depend on x (say, $2r^*$), $N_0(\lambda)$ is expressed explicitly [5, 41]. To mention this, we define a function v on $\mathbf{R}^d \times (0, \infty)$ by

$$v(x, \lambda) = (2\pi)^{-(k+r^*)}\gamma_k \prod_{j=1}^{r^*} b_j(x) \sum_{n_i \in \mathbf{Z}_+} \left(\left(2\lambda - \sum_{i=1}^{r^*}(2n_i + 1)b_i(x)\right)_+\right)^k,$$

where $k = d - 2r^*$, γ_k is the volume of the unit ball in $\mathbf{R}^k$, $f_+ = \max(f, 0)$ and $(f_+)^0$ is the Heaviside function. Then $N_0(\lambda)$ is expressed as

$$N_0(\lambda) = \int_{\mathbf{R}^d} v(x, \lambda)\,dx, \quad \lambda > 0.$$

Combining Theorem 2.1 with the results of Colin de Verdière [5], Ivrii [19–21], Tamura [41], who have studied the asymptotic distribution of the Schrödinger operator H_θ defined by (1.7), we obtain the following corollary that is one of the main results in [28].

COROLLARY 3.2. *Under the assumptions* (H.1)–(H.4), *if* $V(x) = o(\|d\theta\|(x))$ *as* $|x| \to \infty$, *it holds that*

$$\mathrm{Tr}(e^{-tH}) = \mathrm{Tr}(e^{-tH_\theta}) \cdot (1 + o(1)) = J(t) \cdot (1 + o(1))$$

as $t \downarrow 0$. *Moreover, if* $N_0(\lambda)$ *is regularly varying as* $\lambda \uparrow \infty$,

(3.2) $$N(\lambda) = N_0(\lambda) \cdot (1 + o(1))$$

holds as $\lambda \uparrow \infty$.

We should remark here that $N_0(\lambda)$ cannot be expressed in terms of the Hamiltonian like (3.1). This is because, if we have such an expression as (3.1), the volume of the set $\{(x, p); \frac{1}{2}|p - \theta(x)|^2 < \lambda\}$ should give the leading asymptotics, and it is infinite for any $\lambda > 0$. In this sense, if we follow the terminology of Simon [38], the problem treated in Corollary 3.1 is classical and that in Corollary 3.2 is nonclassical.

As an example of nonclassical problems, Robert [35] and Simon [38] have considered the scalar potential given by $V(x_1, x_2) = |x_1|^\alpha |x_2|^\beta$, $\alpha, \beta > 0$, if we work on $\mathbf{R}^2$, and have studied the asymptotic distribution of the eigenvalues for $A = -\frac{1}{2}\Delta + V$. For example, it is shown in [38] that

$$\mathrm{Tr}(e^{-tA}) = c\, t^\mu \cdot (1 + o(1)), \qquad \mu = \frac{\alpha + \beta + 2}{2\alpha},$$

holds as $t \downarrow 0$ if $\alpha < \beta$, where c is a positive constant which is given in terms of α and β.

The Schrödinger operators with degenerate potentials like this example has been studied recently by Tachizawa [39] including more general cases. We also refer to Chapter XI of Edmunds–Evans [9] and references cited therein. Moreover, Mohamed and Nourrigat [30] have considered the Schrödinger operators with magnetic fields in the framework including the example above and has studied the growth rate of $N(\lambda)$ in λ. The methods used in these works are, as in [5], based on the min-max principle.

We end this section with a few examples. In the examples below, θ and V are not necessarily smooth. But I remark here that we can show (2.1) if we slightly modify its proof.

First, we consider the Schrödinger operator given in Helffer [12] as an interesting example which has compact resolvents and generalize it. For the following example, the growth rate of $N(\lambda)$ in λ has been studied in [30].

EXAMPLE 1. Let $d = 2$ and set

$$\theta_1(x_1, x_2) = |x_2|^{k+1}, \quad \theta_2(x_1, x_2) = 0, \quad V(x_1, x_2) = |x_1|^{2l}$$

in (1.7), where $k, l > 0$. Theorem 2.1 implies

$$\mathrm{Tr}(e^{-tH}) = c\, t^{-(1 + 1/k + 1/2l)} \cdot (1 + o(1)),$$

$$c = \frac{1}{2\pi} \int_{\mathbf{R}^2} e^{-|x_1|^{2l}} \frac{(k+1)|x_2|^k/2}{\sinh(k+1)|x_2|^k/2}\, dx_1 dx_2$$

as $t \downarrow 0$. Therefore we get, by the Tauberian theorem,

$$N(\lambda) = \frac{c}{\Gamma(2 + 1/k + 1/2l)}\, \lambda^{1 + 1/k + 1/2l} \cdot (1 + o(1)).$$

For the operators treated in [30, 35, 38] mentioned above, if we add a strong magnetic field, we can apply Corollary 3.2, and the situation is changed. For example, we see the following:

EXAMPLE 2. Let $d = 2$ and set $V(x_1, x_2) = |x_1||x_2|$. Setting $H_1 = -\frac{1}{2}\Delta + V$, we have by [38]
$$\text{Tr}(e^{-tH_1}) = 2\pi^{-1} t^{-2} \log(t^{-1}) \cdot (1 + o(1)).$$

If we consider the Schrödinger operator $H = H_{(\theta,V)}$ which is obtained by adding to H_1 the magnetic field given by $\theta_1(x_1, x_2) = x_1^2 x_2, \theta_2(x_1, x_2) = -x_1 x_2^2$, we get
$$\text{Tr}(e^{-tH}) = c\, t^{-2} \cdot (1 + o(1)), \quad N(\lambda) = \frac{c}{2} \lambda^2 \cdot (1 + o(1)),$$

where the constant c is given by
$$c = (2\pi)^{-1} \int_{\mathbf{R}^2} \frac{(x_1^2 + x_2^2)/2}{\sinh(x_1^2 + x_2^2)/2}\, e^{-|x_1||x_2|}\, dx_1 dx_2.$$

Finally we consider the potential which grows exponentially. Since we have not assumed the upper bound for the growth of the potentials, we can treat also the following:

EXAMPLE 3. Let $d = 2$ and $V(x) = \exp(|x|)$. If the assumptions of Corollary 3.1 are satisfied, it holds that
$$\text{Tr}(e^{-tH}) = \frac{1}{2} t^{-1} \left(\log \frac{1}{t}\right)^2 \cdot (1 + o(1)), \quad N(\lambda) = \frac{1}{2}\lambda(\log \lambda)^2 \cdot (1 + o(1)).$$

4. Some probabilistic considerations

In this section we show that the right-hand side of the formula (2.1) in Theorem 2.1 can be easily guessed by using the Feynman–Kac formula and the Lévy formula for the stochastic area. The difference from the arguments in §1 is in that, because of the magnetic field, a stochastic oscillatory integral appears in the expression of the heat kernel via the integral of a Wiener functional. Our problem is related to the stationary phase methods on the Wiener space, and the relation will be mentioned at the final part.

We denote by $q(t, x, y)$, $t > 0$, $x, y \in \mathbf{R}^d$, the heat kernel for the operator H defined by (1.7). By using the Feynman–Kac formula, its value on the diagonal set is expressed in terms of the functional integral on the Wiener space as follows: using the same notation as those in §1,

(4.1) $\quad q(t, z, z) = (2\pi t)^{-d/2} E_0[e^{\sqrt{-1} F(\omega;t,z)} e^{-t \int_0^1 V(z + \sqrt{t}\omega_s) ds} | \omega_1 = 0],$

where F is defined by
$$F(\omega; t, z) = \sqrt{t} \sum_{j=1}^{d} \int_0^1 \theta_j(z + \sqrt{t}\omega_s) \circ d\omega_s^j$$

and $\circ d\omega_s^j$ denotes the Stratonovich stochastic integral with respect to the Wiener process $\omega = \{\omega_s\}_{s \geq 0}$. For matters related to probability theory, e.g., the Wiener processes and the Itô formula which will be necessary below, we refer to Ikeda–Watanabe [18].

For F, we apply Itô's formula repeatedly. Then we get

$$(4.2) \quad F(\omega;t,z) = \sqrt{t}\sum_{j=1}^{d}\theta_j(z)\omega_1^j + t\sum_{i,j=1}^{d}\partial_i\theta_j(z)\int_0^1 \omega_s^i \circ d\omega_s^j + t^{3/2}L(\omega;t,z),$$

where

$$L(\omega;t,z) = \sum_{i,j,k=1}^{d}\int_0^1 \circ d\omega_s^j \int_0^s \circ d\omega_u^i \int_0^u \partial_k\partial_i\theta_j(z+\sqrt{t}\omega_\tau) \circ d\omega_\tau^k.$$

Under the condition $\omega_1 = 0$, the first term of the right-hand side of (4.2) is equal to 0 and the second term is written, by using the inner product $\langle \cdot, \cdot \rangle$ in $\mathbf{R}^d$, as

$$\sum_{i,j=1}^{d}\partial_i\theta_j(z)\int_0^1 \omega_s^i \circ d\omega_s^j = \frac{1}{2}\int_0^1 \langle B(z)\omega_s, \circ d\omega_s \rangle,$$

where $B(z) = (\partial_i\theta_j(z) - \partial_j\theta_i(z))_{i,j=1,2,\ldots,d}$.

We denote by $\tilde{B}(z)$ the canonical form of the $d \times d$ real skew-symmetric matrix $B(z)$. Using the functions $b_1(z), \ldots, b_{r(z)}(z)$ introduced in §2, the $(2k-1, 2k)$ and the $(2k, 2k-1)$ components are $b_k(z)$ and $-b_k(z)$, respectively. If we let U_z be the transformation matrix (the orthogonal matrix satisfying $U_z B(z) U_z^* = \tilde{B}(z)$), the probability laws of $\{\tilde{\omega}_s\}_{s\geq 0} := \{U_z\omega_s\}_{s\geq 0}$ and $\{\omega_s\}_{s\geq 0}$ coincide by virtue of the rotation invariance of the Wiener measure, and it holds that

$$\sum_{i,j=1}^{d}\partial_i\theta_j(z)\int_0^1 \omega_s^i \circ d\omega_s^j = \frac{1}{2}\int_0^1 \langle \tilde{B}(z)\tilde{\omega}_s, \circ d\tilde{\omega}_s \rangle = \sum_{k=1}^{r(z)} b_k(z)S_1^{2k,2k-1}$$

if $\omega_1 = 0$. Here $\{S_s^{i,j}\}_{s\geq 0}$ defined by

$$S_s^{i,j} = \frac{1}{2}\int_0^s (\tilde{\omega}_u^i \circ d\tilde{\omega}_u^j - \tilde{\omega}_u^j \circ d\tilde{\omega}_u^i)$$

is the Wiener functional called Lévy's stochastic area which can be regarded as a "quadratic function" on the Wiener space.

As to the integral of V in the right-hand side of (4.1), we replace it by $tV(z)$ as in §1. Then we get

$$q(t,z,z) = (2\pi t)^{-d/2} e^{-tV(z)} \prod_{k=1}^{r(z)} E_0[\exp\{\sqrt{-1}tb_k(z)S_1^{2k,2k-1}\}|\tilde{\omega}_1 = 0](1+o(1)).$$

The conditional expectation related to the stochastic area in the right-hand side is well known as the Lévy's formula [18, 26], which implies

$$(4.3) \qquad q(t,z,z) = (2\pi t)^{-d/2} e^{-tV(z)} \prod_{k=1}^{r(z)} \frac{tb_k(z)/2}{\sinh tb_k(z)/2} \cdot (1+o(1))$$

as $t \downarrow 0$. Although we cannot expect that the small order $o(1)$ in the right-hand side of (4.3) is uniform in z, integrating both sides in z formally gives us the assertion of Theorem 2.1. Therefore, if we study the asymptotic behavior of $q(t,z,z)$ as $t \downarrow 0$ at each point $z \in \mathbf{R}^d$ by using the change of the coordinates which depends on z and if we integrate the leading term, we obtain the assertion of Theorem 2.1.

Here we have transformed the functional F by using only Itô's formula to derive (4.3). As a matter of fact, F is the stochastic line integral of the 1-form θ along the path of the Wiener process and, by using Stokes' theorem for the stochastic line integrals [15, 40] and rewriting F in terms of the surface integral with respect to the stochastic area, we can show (4.3).

Now let us return to (4.1). To make the arguments easier, let us assume that $d = 2$ and $V = 0$. There appears a stochastic oscillatory integral in the right-hand side of (4.1). We are interested in the case where $b_1(z) = \|d\theta(z)\| \to \infty$ as $|z| \to \infty$. Moreover, by setting $tb_1(z) = \lambda$, the arguments above imply that F is of the form

$$(4.4) \qquad F(\omega; t, z) = \lambda S_1(\omega) + L(\lambda, \omega)$$

and $L(\lambda, \omega)$ is a negligible term when $\lambda \to \infty$. On the other hand, S_1 is a "quadratic function" on the Wiener space and its "Hessian" is nondegenerate. Furthermore $E_0[\exp(\sqrt{-1}\lambda S_1)|\omega_1 = 0]$ is calculated concretely by Lévy's formula. Therefore we can see an analogue of the stationary phase methods on finite-dimensional spaces in this integral on the Wiener space and our result seems to suggest the stationary phase methods on the Wiener space from another point of view.

The stationary phase methods for the integrals on infinite-dimensional spaces have been already studied by Albeverio–Høegh Krohn [1], Atiyah [2], Ben Arous [4] and so on and, recently, Ikeda–Manabe [16] and Prat [32] have discussed this problem in a similar formulation to that of our problem. But our object is obtained by integrating over $\mathbf{R}^d$ a function which is given by the functional integration and, at present, it is difficult to show that the term related to $L(\lambda, \omega)$ does not contribute to the leading asymptotics when we integrate it over $\mathbf{R}^d$. Furthermore we might not have any general theory which shows that the first term of the right-hand side of (4.4) is dominant for the asymptotic behavior as $\lambda \uparrow \infty$ of the functional integral. Also due to a wide range of its applications, we might hope to make more progress on the study of this problem.

5. On the proof of Theorem 2.1

As is mentioned in the last section, the proof of Theorem 2.1 by using the functional integral representation of the heat kernel in terms of the Feynman–Kac formula and by using the properties of the Wiener processes like that in Ray [33] is not known.

The author [29] has proved Theorem 2.1 by expressing the difference between $q(t, z, z)$, the value of the integral kernel of the semigroup on the diagonal, and the integrand of the right-hand side of (2.1) in terms of the equality of the Duhamel type also used in Tamura [41] and by estimating the error term. In the proof, an explicit expression of the heat kernel for the Schrödinger operator with a uniform magnetic field plays an important role as in Tamura's proof [41], where the eigenfunction expansion is used for the calculation of the trace and for the estimates of the error terms. For this explicit expression, various proofs are known and, in particular, it is proved by using the Feynman–Kac formula and the Lévy formula. Together with the proof of Theorem 2.1, see [29] for details.

ACKNOWLEDGMENT. The author would like to thank Professor Nobuyuki Ikeda for many suggestions and encouragement during his study of the problem mentioned in this article and Professor Kôhei Uchiyama, who kindly read the original version of this article and gave many useful comments.

NOTES ADDED IN TRANSLATION. At the beginning of §3, it is written that we have not succeeded in expressing the asymptotic distribution of the eigenvalues in terms of the potentials. But the author has found a concrete expression, which will be given in "*Asymptotics of the eigenvalue distributions for Schrödinger operators with magnetic field*", to appear in Comm. Partial Differential Equations

References

1. S. Albeverio and R. Høegh-Krohn, *Oscillatory integrals and the method of stationary phase in infinitely many dimensions, with applications to the classical limit of quantum mechanics*. I, Invent. Math. **40** (1977), 59–106.
2. M. F. Atiyah, *Circular symmetry and stationary-phase approximation*, Astérisque **131** (1985), 43–59.
3. J. Avron, I. Herbst and B. Simon, *Schrödinger operators with magnetic fields*. I, *General interactions*, Duke Math. J. **45** (1978), 847–883.
4. G. Ben Arous, *Méthodes de Laplace et de la phase stationire sur l'espace de Wiener*, Stochastics **25** (1988), 125–153.
5. Y. Colin de Verdière, *L'asymptotique de Weyl pour les bouteilles magnétiques*, Comm. Math. Phys. **105** (1986), 327–335.
6. H. L. Cycon, R. G. Froese, W. Kirsch and B. Simon, *Schrödinger operators with application to quantum mechanics and global geometry*, Springer-Verlag, 1987.
7. J. S. de Wet and F. Mandl, *On the asymptotic distribution of eigenvalues*, Proc. Roy. Soc. London Ser. **200** (1950), 572–580.
8. A. Dufresnoy, *Un exemple de champ magnétique dans R^ν*, Duke Math. J. **50** (1983), 729–734.
9. D. E. Edmunds and W. D. Evans, *Spectral theory and differential operators*, Clarendon Press, Oxford, 1987.
10. R. P. Feynman and A. R. Hibbs, *Quantum mechanics and path integrals*, McGraw-Hill, 1965.
11. D. Fujiwara and K. Asada, *Construction of the fundamental solutions for Schrödinger equations*, Sûgaku **33** (1981), 97–119. (Japanese)
12. B. Hellfer, *Semi-classical analysis for the Schödinger operator and applications*, Lecture Notes in Math., vol. 1336, Springer, 1988.
13. B. Helffer and A. Mohamed, *Caractérisation du spectre essentiel de l'opérateur de Schrödinger avec un champ magnétique*, Ann. Inst. Fourier (Grenoble) **38** (1988), 123–125.
14. T. Ichinose, *On the path integrals for the Dirac equations*, Sûgaku **42** (1990), 219–230. (Japanese)
15. N. Ikeda and S. Manabe, *Integral of differential forms along the paths of diffusion processes*, Publ. RIMS Kyoto Univ. **15** (1979), 827–852.
16. _____, *Asymptotic formulas for stochastic integrals*, Asymptotic Problems in Probability Theory, Longman, New York (to appear).
17. N. Ikeda and H. Matsumoto, *Short time asymptotics for the traces of heat kernels of Schrödinger operators with magnetic field*, Bull. Sci. Math. **116** (1992), 53–66.
18. N. Ikeda and S. Watanabe, *Stochastic differential equations and diffusion processes*, 2nd ed., Kodansha, Tokyo, 1989.
19. V. Ivrii, *Estimations pour le nombre de valeurs propres négatives de l'opérateur de Schrödinger avec potentiels singuliers*, C. R. Acad. Sci. Paris **302** (1986), 467–470, 491–494, 535–538.
20. _____, *Estimates for the number of negative eigenvalues of the Schrödinger operator with a strong magnetic field*, Soviet Math. Dokl. **36** (1988), 561–564.
21. _____, *Sharp spectral asymptotics for the two-dimensional Schrödinger operator with a strong magnetic field*, Soviet Math. Dokl. **39** (1989), 437–441.
22. A. Iwatsuka, *Magnetic Schrödinger operators with compact resolvent*, J. Math. Kyoto Univ. **26** (1986), 357–374.
23. M. Kac, *Probability and related topics in physical sciences*, Interscience, 1959.
24. _____, *Can you hear the shape of a drum*, Amer. Math. Monthly **73** (1966), 1–23.
25. _____, *Integration in function spaces and some of its applications*, Lezioni Fermiane, Pisa, 1980.
26. P. Lévy, *Wiener's random function, and other Laplacian random functions*, 2nd Sympos. Prob. and Stat., Univ. of California Press, Berkeley, 1950, pp. 171–186.
27. H. Matsumoto, *The short time asymptotics of the traces of the heat kernels for the magnetic Schrödinger operators*, J. Math. Soc. Japan **42** (1990), 677–689.
28. _____, *Classical and non-classical eigenvalue asymptotics for magnetic Schrödinger operators*, J. Funct. Anal. **95** (1991), 460–482.

29. H. Matsumoto, *The asymptotic distributions of the eigenvalues for the Schrödinger operators with magnetic fields*, Asymptotic Problems in Probability Theory (K. D. Elworthy and N. Ikeda, eds.), Longman, New York, 1993, pp. 169–193.
30. A. Mohamed and J. Nourrigat, *Encadrement du $N(\lambda)$ pour un opérateur de Schrödinger avec un champ magnétique et un potentiel électrique*, J. Math. Pure Appl. **70** (1991), 87–99.
31. K. Odencrantz, *The effect of a magnetic field on asymptotics of the trace of the heat kernel*, J. Funct. Anal. **79** (1988), 398–422.
32. J-J. Prat, *Équation de Schrödinger: Analyticité transverse de la densité de la loi d'une fonctionnelle additive*, Bull. Sci. Math. (2) **115** (1991), 133–176.
33. D. B. Ray, *On spectra of second order differential operators*, Trans. Amer. Math. Soc. **77** (1954), 299–321.
34. M. Reed and B. Simon, *Methods of modern mathematical physics*, II and IV, Academic Press, London, 1975.
35. D. Robert, *Comportment asymptotique des valeurs propres d'opérateurs du type Schrödinger à potentiel «dégénéré»*, J. Math. Pures Appl. (9) **61** (1982), 275–300.
36. G. V. Rosenbljum, *Asymptotics of the eigenvalues of the Schrödinger operators*, Math. USSR-Sb. **22** (1974), 349–371.
37. B. Simon, *Functional integration and quantum physics*, Academic Press, 1979.
38. _____, *Non-classical eigenvalue asymptotics*, J. Funct. Anal. **53** (1983), 84–98.
39. K. Tachizawa, *Eigenvalue asymptotics of Schrödinger operators with positive potentials*, (preprint).
40. Y. Takahashi and S. Watanabe, *The probability functionals (Onsager-Machlup functions) of diffusion processes*, Stochastic Integrals, Lecture Notes in Math., vol. 851 (D. Williams, ed.), Springer, 1981, pp. 433–463.
41. H. Tamura, *Asymptotic distribution of eigenvalues for Schrödinger operators with magnetic fields*, Nagoya Math. J. **105** (1987), 40–69.
42. E. C. Titchmarsh, *On the asymptotic distribution of eigenvalues*, Quart. J. Math. Oxford Ser. (2) **5** (1954), 228–240.
43. _____, *Eigenfunction expansions*, vols. I, II, Oxford Univ. Press, London and New York, 1962.
44. S. Watanabe, *Analysis of Wiener functionals (Malliavin calculus) and its application to heat kernels*, Ann. Probab. **15** (1987), 1–39.

DEPARTMENT OF MATHEMATICS, FACULTY OF GENERAL EDUCATION, GIFU UNIVERSITY, YANAGIDO, GIFU 501-11, JAPAN

Translated by HIROYUKI MATSUMOTO

Recursive Estimation of Nonparametric Probability Density Functions

Eiichi Isogai

0. Introduction

The approach to estimating parameters on the basis of a sample from a distribution with a known functional form including unknown parameters, such as a normal distribution with unknown mean and variance, is called a parametric method. In contrast, the approach in the case where the distribution has an unknown functional form is called a nonparametric method. Nonparametric methods are very powerful when the functional forms of distributions are not known. Since Rosenblatt [66] considered the problem of estimating nonparametric probability density functions by using the kernel method, this problem has been studied by a large number of investigators, for example, by Parzen [62]. The reason is that this problem is an interesting one in statistical inference and is important in applications, such as pattern recognition ([26], [84], etc.). On the other hand, other methods of estimation, the histogram method, the orthogonal series method, and so on have been proposed and investigated ([27], [93], etc.).

The estimators proposed by Rosenblatt and Parzen are not recursive. In other words, when more observations become available the estimators have to be computed from the beginning. Moreover, all data have to be stored to compute the estimators. To improve these points Wolverton and Wagner [94] and Yamoto [95] have proposed recursive estimators. In this exposition, we shall explain the problem of estimating nonparametric probability density functions, mainly recursive estimators obtained by the kernel method. This problem of estimation is summarized in the books [17], [21], [58], [64], [74], [77], [85], and so on. Furthermore, we shall mention the problem of sequential estimation with random sample sizes. Finally, as a related problem we shall briefly describe the problem of nonparametric regression estimation.

1. Problem and recursive estimators

Let $\mathscr{B}_p$ be the field of p-dimensional Borel sets, and let X and $\{X_n, n \geq 1\}$ be a sequence of p-dimensional random vectors defined on a probability space $(\Omega, \mathscr{F}, P)$, having a common distribution λ_x on $\mathscr{B}_p$. Here suppose that λ_x has a probability density function (abbreviated to p.d.f.) f with an unknown functional form with

---

1991 *Mathematics Subject Classification.* Primary 62G05.
This article originally appeared in Japanese in Sûgaku **45** (1) (1993), 27–41.

respect to p-dimensional Lebesgue measure μ, and that the p.d.f. f belongs to a class $\mathcal{S}$ of p-dimensional probability density functions. The problem of estimating probability density functions is to construct an estimator $f_n(x) \equiv f_n(x; X^n)$ of $f(x)$ ($x \in R^p$) based on a sample $X^n = (X_1, \ldots, X_n) \in R^{pn}$. The estimator $f_n(x)$ is called a probability density estimator if $f_n(x)$ is a probability density function for each fixed $n(\geq 1)$ and X^n. We say that $\{f_n(x), n \geq 1\}$ is recursive if for each n and $j = 1, \ldots, n$ there exists a measurable function Ψ_n such that $f_n(x) = \Psi_n(f_{n-1}(x), x, X_n)$. When $g_{n,j}: R^p \times R^p \to R$ is a measurable function for each n and $j = 1, \ldots, n$, the estimator $f_n(x)$ of the form $f_n(x) = \sum_{j=1}^{n} g_{n,j}(x; X_j)$ is called the kernel estimator.

Several recursive kernel probability density estimators have been proposed and their statistical properties investigated [92]. Now let a real-valued Borel measurable function (called the kernel) $K(y) = K(y_1, \ldots, y_p)$ ($y = (y_1, \ldots, y_p) \in R^p$) and a sequence of positive numbers (called bandwidths) $\{h_n, n \geq 1\}$ be given. In what follows, we shall enumerate the kernel estimators of f by using the K and $\{h_n, n \geq 1\}$ proposed so far. Rosenblatt [66], Parzen [62], and Cacoullos [7] have proposed the following nonrecursive estimator in the one-dimensional case ($p = 1$) and the multidimensional case ($p \geq 1$). We call this estimator the Rosenblatt-Parzen estimator, abbreviated RPE. Similar abbreviations are used for other estimators below.

(RPE) $$f_n(x) = \frac{1}{nh_n^p} \sum_{j=1}^{n} K\left(\frac{x - X_j}{h_n}\right).$$

Now modifying the nonrecursive RPE, Wolverton and Wagner [94] and Yamato [95] have proposed the following recursive WWYE.

(WWY) $$f_n(x) = \frac{1}{n} \sum_{j=1}^{n} K_j(x - X_j), \quad \text{where } K_j(y) = \frac{1}{h_j^p} K\left(\frac{y}{h_j}\right).$$

The WWYE can be computed recursively as follows:

$$f_n(x) = \left(1 - \frac{1}{n}\right) f_{n-1}(x) + \frac{1}{n} K_n(x - X_n).$$

Other recursive estimators, the DE ([14]), the BE ([4]), the WDE ([91]), and the IE ([37]), are proposed as follows:

(DE) $$f_n(x) = \frac{1}{b_n} \sum_{j=1}^{n} H(h_j) K\left(\frac{x - X_j}{h_j}\right),$$

where $b_n = \sum_{j=1}^{n} h_j^p H(h_j)$ and $H: (0, \infty) \to (0, \infty)$ is an arbitrary function. In particular, the DE with $H(y) \equiv 1$ is called DE(1).

(BE) $$f_n(x) = \frac{1}{d_n} \sum_{j=1}^{n} \frac{h_j^p}{d_j} \sum_{i=1}^{j} K\left(\frac{x - X_j}{h_i}\right), \quad \text{where } d_n = \sum_{j=1}^{n} h_j^p.$$

(WDE) $$f_n(x) = \frac{1}{nh_n^{p/2}} \sum_{j=1}^{n} \frac{1}{h_j^{p/2}} K\left(\frac{x - X_j}{h_j}\right).$$

(IE) $$\begin{aligned} f_0(x) &\equiv K(x), \\ f_n(x) &\equiv (1 - a_n) f_{n-1}(x) + a_n K_n(x - X_n) \quad (n \geq 1), \end{aligned}$$

where $\{a_n, n \geq 1\}$ is a sequence of positive numbers satisfying $a_n \leq 1$, $\lim_{n\to\infty} a_n = 0$ and $\sum_{n=1}^{\infty} a_n = \infty$. The IE is a slight modification of the estimator given by Watanabe [88]. In particular, the IE with $a_n = a/n$ for any fixed $0 < a \leq 1$ is called IE(a). Then the IE(1) becomes the WWYE. Moreover, the DE with $H(y) = y^{-p}$ becomes the WWYE. If K is a probability density function, then the RPE, the WWYE, the DE, the BE, and the IE are also probability density estimators. In this exposition we shall explain the statistical properties of the recursive estimators by using kernels. With regard to the RPE, a large number of interesting results on its statistical properties are known, but we do not mention them in detail (see for example, [6], [18], [19], [67]).

To close this section, we summarize the notation used throughout this exposition. $E(\cdot)$ and $V(\cdot)$ denote the expectation and the variance with respect to a given p.d.f. f, respectively. "$\to$ in prob." and "$\to$ a.s." stand for convergence in probability and almost sure convergence, respectively. Set

$$\gamma_0 = \gamma_1 = 1, \qquad \gamma_n = \prod_{j=2}^{n}(1 - a_j) \quad (n \geq 2).$$

Then the IE can be written as follows:

$$(1.1) \qquad f_n(x) = \gamma_n \sum_{j=0}^{n} a_j \gamma_j^{-1} K_j(x - X_j) \qquad (n \geq 0),$$

where $a_0 = 1$, $h_0 = h_1$, $X_0 = 0$, $K_0(x) = K(x)$. Let $L_1 = L_1(R^p, \mathscr{B}_p, \mu)$ denote the set of all integrable real-valued functions, and let $\|\cdot\|$ stand for the Euclidean norm on R^p. Throughout this exposition all kernels K are always assumed to be bounded integrable real-valued Borel measurable functions defined on R^p satisfying

$$(K1) \qquad \|y\|^p |K(y)| \to 0 \quad (\|y\| \to \infty) \quad \text{and} \quad \int K(y)\,dy = 1,$$

where the integral is taken over R^p and we omit the domain. Let the radial majorant φ of K be defined by $\varphi(x) = \sup_{\|y\| \geq \|x\|} |K(y)|$, and let the sequence of bandwidths $\{h_n, n \geq 1\}$ converge to zero as n tends to infinity.

2. Criteria for good estimators

So far, some criteria for comparing various types of estimators have been introduced ([90]). Here we shall give some of them. In what follows, let $\{f_n(x) \equiv f_n(x; X^n)\}$ be a given sequence of estimators. We shall consider convergence as $n \to \infty$. The estimator f_n (more precisely, the sequence $\{f_n\}$) is said to be pointwise consistent in the sense of the mean squared error (abbreviated to MSE) if the MSE of $f_n(x)$, $\text{MSE}(f_n(x)) \equiv E(f_n(x) - f(x))^2$, converges to zero for every x and every $f \in \mathscr{S}$. The estimator f_n is said to be uniformly consistent in the sense of the MSE if $\sup_x \text{MSE}(f_n(x))$ tends to zero. The estimator f_n is called weakly (strongly) pointwise consistent if $f_n(x)$ converges to $f(x)$ in probability (almost surely) for every x and every $f \in \mathscr{S}$. The estimator f_n is called weakly (strongly) uniformly consistent if this convergence is uniform in x. Note that pointwise consistency, in the sense of the MSE, implies weak pointwise consistency. Moreover, strong pointwise consistency is preferable to weak pointwise consistency, and uniform consistency is preferable to pointwise consistency. For a positive number r, the estimator f_n is said to be weakly (strongly) consistent in L_r if for every $f \in \mathcal{S}$ $\int |f_n(x) - f(x)|^r \, dx$ converges to zero in probability (almost surely). In particular, when the mean integrated squared error (abbreviated to MISE) of f_n $E[\int |f_n(x) - f(x)|^2 \, dx]$ converges to zero, the estimator f_n is called consistent in the sense of the MISE.

3. Independent observations

3.1. Asymptotic unbiasedness. In §3, we suppose that $X_1, \ldots, X_n$ are independent identically distributed p-dimensional random vectors with a common p.d.f. f. It is known that there exists no probability density estimator f_n satisfying $Ef_n(x) = f(x)$ for all x and all f ([66], [75]). Therefore, one of our interesting subjects is whether or not a given sequence of estimators has asymptotic unbiasedness, that is, whether or not $\lim_{n \to \infty} Ef_n(x) = f(x)$ holds. The following lemma plays a fundamental role in proving asymptotic properties [7].

LEMMA 3.1. *For $g \in L_1$ and a positive number h let*

$$K_h(y) = \frac{1}{h^p} K\left(\frac{y}{h}\right)$$

and

$$g * K_h(x) = \int g(x-y) K_h(y) \, dy = \int K_h(x-y) g(y) \, dy,$$

where $*$ denotes the convolution operator. Then it holds that $\lim_{h \to +0} g * K_h(x) = g(x) \int K(y) \, dy$ for every point x of continuity of g.

Here $\int K(y) \, dy = 1$ is not necessarily satisfied in the above lemma. Due to Stein [79, pages 62–63], the following result is obtained when the continuity of g is not assumed.

LEMMA 3.2. *Suppose that $g \in L_1$ and that K has a radial majorant $\varphi \in L_1$. Then it follows that $\lim_{h \to +0} g * K_h(x) = g(x)$ for almost all x.*

Here one of the sufficient conditions for $\varphi \in L_1$ is that the K assumed in §1 satisfies $K(x) = O(\|x\|^{-p-\varepsilon})$ ($\|x\| \to \infty$) for some $\varepsilon > 0$.

We shall now consider asymptotic unbiasedness. Using Lemma 3.1 and the Toeplitz lemma ([53, page 250]) we have the following result.

THEOREM 3.1. *Let x be any point of continuity of any f. Then the RPE and the IE are asymptotically unbiased. If $\lim_{n \to \infty} b_n = \infty$, then the DE is asymptotically unbiased. Further, the BE is asymptotically unbiased, provided that $\lim_{n \to \infty} d_n = \infty$.*

3.2. Consistency. Since $\text{MSE}(f_n(x)) = V(f_n(x)) + B_n^2(x)$ ($B_n(x) = Ef_n(x) - f(x)$) by using Lemma 3.1 and Theorem 3.1, we have the following result on pointwise consistency in the sense of the MSE.

THEOREM 3.2. *Let x be any point of continuity of any f. Then the RPE is pointwise consistent in the sense of the MSE if $\lim_{n \to \infty} nh_n^p = \infty$. If $\lim_{n \to \infty} b_n = \infty$ and $\lim_{n \to \infty} b_n^{-2} \sum_{j=1}^n h_j^p H^2(h_j) = 0$, then the DE is pointwise consistent in the sense of the MSE. Moreover, the IE is pointwise consistent in the sense of the MSE, provided that $\lim_{n \to \infty} a_h h_n^{-p} = 0$.*

By using the relation that $f_n(x) - f(x) = (f_n(x) - Ef_n(x)) + B_n(x)$ and Theorem 3.1, we obtain the following result on strong pointwise consistency ([37], [64]).

THEOREM 3.3. *Let x be any point of continuity of any f. Then the RPE is strongly pointwise consistent and pointwise consistent in the sense of the MSE, provided that $\sum_{n=1}^{\infty} \exp(-cnh_n^p) < \infty$ for every constant $c > 0$. If $\sum_{n=1}^{\infty} a_n^2 h_n^{-p} < \infty$, then the IE is strongly pointwise consistent and pointwise consistent in the sense of the MSE.*

When the continuity of f is not assumed, we have the following result by using Lemma 3.2, [21].

THEOREM 3.4. *Let f be any p.d.f., and suppose that K is nonnegative on R^p with a radial majorant $\varphi \in L_1$. Then the WWYE is weakly pointwise consistent for almost all x if $\lim_{n \to \infty} nh_n^p = \infty$. If also $\lim_{n \to \infty} (nh_n^p / \log_2 n) = \infty$, then the WWYE is strongly pointwise consistent for almost all x, where $\log_2 n = \log\log n$.*

We note that Roussas [70] has discussed the exact rates of almost sure pointwise convergence of the WWYE to p.d.f. f.

By using the Fourier transform of kernel, we can show the strong uniform consistency of the IE, [37]. Set

$$k(t) = \int \exp\left(i \sum_{j=1}^{p} t_j y_j\right) K(y)\, dy, \qquad i^2 = -1, \ A_n = \gamma_n^2 \sum_{j=1}^{n} a_j^2 \gamma_j^{-2}.$$

THEOREM 3.5. *Let f be any uniformly continuous p.d.f. on R^p. Suppose that K is continuous on R^p and that $|k(t)| \in L_1$ is nonincreasing on $R(u) = \{t = qu; q > 0\}$ for all $u \neq 0 \in R^p$; that is, $|k(t_1)| \geq |k(t_2)|$ for $t_1, t_2 \in R(u)$ with $\|t_1\| \leq \|t_2\|$. In addition, a sequence of bandwidths $\{h_n\}$ is nondecreasing and satisfies*

$$\lim_{n \to \infty} \frac{h_n}{h_{n+1}} = 1, \qquad \sum_{n=1}^{\infty} (a_n h_n^{-p})^2 < \infty, \qquad \sum_{n=1}^{\infty} \frac{A_n}{h_n^{2p-1}} \left|\frac{1}{h_{n+1}} - \frac{1}{h_n}\right| < \infty.$$

Then the IE is strongly uniformly consistent and uniformly consistent in the sense of the MSE.

REMARK 3.1. In the one-dimensional case, under certain conditions on K and $\{h_n\}$, a necessary and sufficient condition for the RPE to be strongly uniformly consistent is that f is uniformly continuous, [73].

Let $I(A)$ denote the indicator function of A. Then the result on the consistency in L_1 of the DE(1) is obtained below, [21].

THEOREM 3.6. *Let K be nonnegative on R^p with a radial majorant $\varphi \in L_1$, and let $\{h_n, n \geq 1\}$ be a sequence of positive numbers. Then the following statements are equivalent*:
  (i) *The DE(1) is strongly pointwise consistent for all f and almost all x*;
  (ii) *The DE(1) is weakly pointwise consistent for some f and almost all x*;
  (iii) $\lim_{n \to \infty} b_n = \infty$ *and* $\lim_{n \to \infty} \sum_{j=1}^{n} h_j^p I(h_j > \varepsilon)/b_n = 0$ *for all* $\varepsilon > 0$.
  (iv) *The DE(1) is strongly consistent in L_1 for all f*;
  (v) *The DE(1) is weakly consistent in L_1 for some f*.

Consistency in L_1 for the WWYE is also discussed in [16].

3.3. Asymptotic variance.
We shall compare the estimators from the viewpoint of asymptotic variance. Set

$$(3.1) \qquad I_0 = f(x) \int K^2(y)\, dy.$$

Using the relation

$$V\left(K\left(\frac{x-X}{h}\right)\right) = h^p f * (K^2)_h(x) - h^{2p}(f * K_h(x))^2$$

and Lemma 3.1, we have the following result on the asymptotic variance ([4], [35], etc.).

THEOREM 3.7. *Let x be any point of continuity of any f.*
(i) *For the RPE and the WDE, it holds that*

$$\lim_{n \to \infty} nh_n^p V(f_n(x)) = I_0.$$

(ii) *If there exists a positive number α such that*

$$\lim_{n \to \infty} \frac{nh_n^p}{b_n^2} = 0 \quad \text{and} \quad \lim_{n \to \infty} \frac{nh_n^p}{b_n^2} \sum_{j=1}^n h_j^p H^2(h_j) = \alpha,$$

then for the DE it holds that

$$\lim_{n \to \infty} nh_n^p V(f_n(x)) = \alpha I_0.$$

(iii) *If $\{h_n\}$ is a monotone decreasing sequence and if for some $a \in (1/2, 1]$ there exists a positive number β such that*

$$\lim_{n \to \infty} n^{1-2a} h_n^p \sum_{j=1}^n j^{2(a-1)} h_j^{-p} = \beta,$$

then for the IE(a) it follows that

$$\lim_{n \to \infty} nh_n^p V(f_n(x)) = a^2 \beta I_0.$$

REMARK 3.2. When $H(y) \equiv 1$, $(nh_n^p/b_n^2) \sum_{j=1}^n h_j^p H^2(h_j)$ obtains its minimum, so that α is also at its minimum.

We shall now introduce the notation for the comparison of the estimators. Let $\{f_n\}$ and $\{g_n\}$ be two sequences of estimators. We use the notation $f_n <_1 g_n$ ($f_n <_2 g_n$) when there exist some constant $\eta \in (0, 1)$ and a positive integer N such that $V(f_n) \leq \eta V(g_n)$ (MSE(f_n) $\leq \eta$MSE(g_n)) for all $n \geq N$. Thus, if $f_n <_1 g_n$ ($f_n <_2 g_n$), then $\{f_n\}$ is better than $\{g_n\}$ from the viewpoint of asymptotic variance (asymptotic MSE). Here in the case where $h_n = n^{-r/p}$ ($0 < r < 1/2$) we shall compare the estimators from the point of view of asymptotic variance. In this case, putting $H(y) \equiv 1$, the conditions of Theorem 3.7 with $\alpha = 1 - r$ and $\beta = (2a + r - 1)^{-1}$ are fulfilled. Therefore, for every point x of continuity of f with $f(x) > 0$ it holds that

$$\text{DE}(1), \text{IE}(1 - r) <_1 \text{WWYE} <_1 \text{WDE, RPE}.$$

3.4. Asymptotic expansions of the MSE.
In this section, in the one-dimensional case ($p = 1$) we shall give each asymptotic expansion of the MSE corresponding

to the RPE, the DE(1) and the IE(a), and compare these estimators from the point of view of asymptotic MSE. A discussion similar to this was presented, for example, in [15]. In what follows, suppose that K satisfies

(K2) $\qquad K(y) = K(-y) \ (y \in R), \qquad \int y^2 |K(y)| \, dy < \infty$

and that f is twice continuously differentiable on R and f'' is bounded. Then by using the Taylor theorem, we have

$$f * K_h(x) = f(x) + I_1 h^2 + o(h^2) \qquad (h \to 0),$$
$$f * (K^2)_h(x) = I_0 + I_2 h^2 + o(h^2) \qquad (h \to 0),$$

where

$$I_m = \frac{f''(x)}{2} \int y^2 K^m(y) \, dy \qquad (m = 1, 2)$$

and I_0 is as given by (3.1). From these relations we can obtain the following result on the asymptotic expansions of the bias $B_n(x)$ of the estimators $f_n(x)$.

THEOREM 3.8. *Let f_n be one of the RPE, the DE(1), and the IE(a). Then it holds that*

$$B_n(x) = I_1 \xi_n + o(\xi_n) \qquad (n \to \infty),$$

where

(i) $\xi_n = h_n^2$ *for the RPE*,
(ii) $\xi_n = b_n^{-1} \sum_{j=1}^n h_j^3$ *for the DE(1), provided that* $\sum_{n=1}^\infty h_n^3 = \infty$,
(iii) $\xi_n = a n^{-a} \sum_{j=1}^n j^{a-1} h_j^2$ *for the IE(a), provided that* $\sum_{n=1}^\infty n^{a-1} h_n^2 = \infty$.

We consider here the case where $h_n = n^{-r}$ ($0 < r < 1/3$). Then from Theorems 3.7 and 3.8 we can find the asymptotic expansions of the MSE.

THEOREM 3.9. *Let $c_n = o(n^{r-1} + n^{-4r})$. Then the following hold*:

$$\text{MSE(RPE)} = I_0 n^{r-1} + I_1^2 n^{-4r} + c_n,$$
$$\text{MSE(DE(1))} = (1-r) I_0 n^{r-1} + \left(\frac{1-r}{1-3r} \right)^2 I_1^2 n^{-4r} + c_n,$$
$$\text{MSE(WWYE)} = \frac{I_0}{1+r} n^{r-1} + \frac{I_1^2}{(1-2r)^2} n^{-4r} + c_n,$$
$$\text{MSE(IE}(1-r)) = (1-r) I_0 n^{r-1} + \left(\frac{1-r}{1-3r} \right)^2 I_1^2 n^{-4r} + c_n.$$

REMARK 3.3. (i) Each right-hand side of the MSE's is asymptotically minimized by $r = 1/5$. The asymptotically optimal choice of bandwidth is also considered (see for example, [81]).

(ii) There are many discussions on the asymptotically optimal choice of kernel (see for instance, [11]).

Now by using Theorem 3.9, we shall compare the estimators from the point of view of asymptotic MSE. In case $0 < r < 1/5$, the relation RPE $<_2$ WWYE $<_2$ DE(1), IE($1-r$) holds for each x with $f''(x) \neq 0$. In the case $1/5 < r < 1/3$, the relation DE(1), IE($1-r$) $<_2$ WWYE $<_2$ RPE holds for each x with $f(x) > 0$. In case $r = 1/5$, the comparison of the estimators cannot necessarily be done for this h_n.

However, by using $h_n = (I_0/4I_1^2)^{1/5}n^{-1/5}$ and K_0 given by (3.2) in §3.5, the relation RPE $<_2$ WWYE $<_2$ DE(1) holds for each x with $f(x) > 0$ and $f''(x) \neq 0$, [92].

From the viewpoint of the MISE, each MISE corresponding to the kernel method and the histogram method is given as follows: suppose that f is twice continuously differentiable on R and that f'' is bounded and $f, f', f'' \in L_2$. Assume that K is a p.d.f. and satisfies $K(y) = K(-y)$ $(y \in R)$ and $\int y^2 K(y)\,dy = 1$. Let f_n be the RPE. Now setting $L = \int K^2(y)\,dy$ and $M = \int (f''(y))^2\,dy$, we consider the asymptotically optimal $h_n = (L/M)^{1/5}n^{-1/5}$. Then we have $\text{MISE}(f_n) = \{5M^{1/5}L^{4/5}/4\}n^{-4/5} + o(n^{-4/5})$. The kernel which minimizes this right-hand side is given by (3.2), [23]. On the other hand, let $Q = \{C_j\}$ be a partition of R $(\bigcup_j C_j = R;\ C_i \cap C_j = \emptyset$ whenever $i \neq j)$, and let $\mathscr{P}$ be the set of all partitions of R. For some partition $Q = \{C_j\}$, the histogram on $x \in C_j$ based on a sample $X_1, \ldots, X_n$ is defined by $H_n(x|Q) = n^{-1}\sum_{i=1}^n I\{X_i \in C_j\}/|C_j|$, where $|C_j|$ stands for the width of cell C_j. Let $\text{MISE}(n, Q)$ denote the MISE of the histogram $H_n(x|Q)$. Then we obtain

$$\inf_{Q \in \mathscr{P}} \text{MISE}(n, Q) = \left\{\left(\frac{3}{2}\right)\left(\frac{1}{6}\right)^{1/3}\int |f'(y)f(y)|^{2/3}\,dy\right\}n^{-2/3} + o(n^{-2/3})$$

([48], [49]).

3.5. Asymptotic normality. For simplicity we shall consider the asymptotic normality of the IE(a) for $h_n = n^{-r/p}$ $(p/(p+2) < r < 1, 2a + r > 1)$ (for a more general case, see [39]). Let K satisfy

(K3) $$\int \|y\|\,|K(y)|\,dy < \infty.$$

THEOREM 3.10. *Suppose that f is partially differentable on R^p and that the partial derivatives $\partial f(x)/\partial x_j$ $(j = 1, \ldots, p)$ are bounded. Let $q \geq 1$ be any given integer, and let $x_1, \ldots, x_q$ be any distinct points with $f(x_j) > 0$ $(j = 1, \ldots, q)$. Then $(nh_n^p)^{1/2}(f_n(x_1) - f(x_1), \ldots, f_n(x_q) - f(x_q))$ converges in law to the q-dimensional normal distribution with mean vector $\mathbf{0}$ and covariance matrix Γ, where*

$$\Gamma = \frac{a^2}{2a+r-1}\int K^2(y)\,dy \begin{pmatrix} f(x_1) & & 0 \\ & \ddots & \\ 0 & & f(x_q) \end{pmatrix}.$$

REMARK 3.4. It can be seen that $(nh_n^p)^{1/2}(f_n(x_j) - f(x_j))$ $(j = 1, \ldots, q)$ are asymptotically independent and normal with mean 0 and variance

$$\frac{a^2}{2a+r-1}f(x_j)\int K^2(y)\,dy.$$

Recently, Isogai ([47]) has discussed a Berry-Esseen type bound for the WWYE.

Let us give some examples of kernels. For $K_0(t)$ $(t \in R)$ below, put $K(y) = \prod_{j=1}^p K_0(y_j)$. Then K satisfies (K1), (K2), and (K3). In particular, each kernel K corresponding to K_0 in (3.3) below fulfills the conditions of Theorem 3.5 as well.

(3.2)
$$K_0(t) = \frac{1}{2}I(|t| \leq 1), \qquad K_0(t) = (1-|t|)I(|t| \leq 1),$$
$$K_0(t) = \frac{3}{4\sqrt{5}}\left(1 - \frac{t^2}{5}\right)I(|t| \leq \sqrt{5}),$$

$$(3.3) \qquad K_0(t) = (2\pi)^{-1/2} \exp\left(-\frac{t^2}{2}\right), \qquad K_0(t) = \frac{1}{2}\exp(-|t|).$$

4. Dependent observations

Suppose that $\{X_t; t = 0, \pm 1, \pm 2, \dots\}$ is a strictly stationary process taking values in R^p with a p.d.f. f. Let $\mathcal{M}_a^b$ ($a \leq b$) be the σ-field generated by $X_a, \dots, X_b$, and let $L_2(\mathcal{M}_a^b)$ denote the set of all $\mathcal{M}_a^b$-measurable random variables U such that $E(U^2) < \infty$. Here, as dependence conditions, we shall define some mixing conditions.

DEFINITION 4.1. (i) $\{X_t\}$ is said to satisfy the strong mixing condition if

$$\alpha(n) = \sup_{A \in \mathcal{M}_{-\infty}^0, B \in \mathcal{M}_n^\infty} |P(A \cap B) - P(A)P(B)| \to 0 \qquad (n \to \infty).$$

(ii) $\{X_t\}$ is said to satisfy the absolute regularity condition if

$$\beta(n) = E\left\{\sup_{A \in \mathcal{M}_n^\infty} |P(A|\mathcal{M}_{-\infty}^0) - P(A)|\right\} \to 0 \qquad (n \to \infty),$$

where $P(\cdot|\mathcal{M}_{-\infty}^0)$ denotes a conditional probability.

(iii) $\{X_t\}$ is said to satisfy the complete regularity condition if

$$\gamma(n) = \sup \frac{|E(\xi\eta) - E(\xi)E(\eta)|}{(V(\xi)V(\eta))^{1/2}} \to 0 \qquad (n \to \infty),$$

where the supremum is taken over all $\xi \in L_2(\mathcal{M}_{-\infty}^0)$ and $\eta \in L_2(\mathcal{M}_n^\infty)$.

If the absolute regularity condition or the complete regularity condition is satisfied, then the strong mixing condition is also satisfied. Further mixing conditions are defined and their properties are investigated (for example, [96]). Now, for the WWYE, Györfi [32] showed strong consistency in L_2. For the WWYE and the WDE, Masry [55] gave pointwise consistency in the sense of the MSE and asymptotic normality under the strong mixing and complete regularity conditions. For the RPE and the WWYE, Takahata [83] discussed the rates of convergence of strong uniform consistency under the absolute regularity condition. In the multidimensional case, for the WWYE $f_n(x)$, Masry [54] presented the following result on strong pointwise consistency and its rate of convergence under the strong mixing condition.

THEOREM 4.1. *Suppose that K has a radial majorant $\varphi \in L_1$.*
(i) *If for some $\delta > 0$ and some $r > 2$*

$$\sum_{n=1}^\infty (\log n)(\log_2 n)^{1+\delta}[\alpha(n)]^{1-(2/r)} \sum_{j=n}^\infty \frac{1}{j^2 h_j^{2p(1-1/r)}} < \infty$$

and

$$\sum_{n=1}^\infty \frac{1}{n^2 h_n^{2p(1-1/r)}} < \infty,$$

then the WWYE is strongly pointwise consistent for almost all $x \in R^p$.

(ii) *Suppose that f is twice partially differentiable and that all the second order partial derivatives are bounded and continuous. Moreover, assume that K satisfies*

$$\int y_j K(y)\, dy = 0 \;(j = 1,\ldots,p) \quad \text{and} \quad \int \|y\|^2 |K(y)|\, dy < \infty,$$

and that for some $r > 2$

$$\sum_{n=1}^{\infty} (\log n)(\log_2 n)[\alpha(n)]^{1-(2/r)} < \infty.$$

Then setting $h_n = n^{-1/\gamma}$ ($\gamma = 4 + 2p(1 - 1/r)$) we have that for every x and every $\delta > 0$

$$\left\{ \frac{n^{4/\gamma}}{(\log n)(\log_2 n)^{1+\delta}} \right\}^{1/2} (f_n(x) - f(x)) \to 0 \quad a.s.\; (n \to \infty).$$

For the WWYE, Masry and Györfi [56] discussed the rate of convergence of strong pointwise consistency under the complete regularity condition. Tran [86] improved the result of [54] in the one-dimensional case. Besides, in the multidimensional case, under another dependence condition Tran [87] provided the rates of convergence of strong uniform consistency and the asymptotic normality for the WWYE. The discussion on Markov processes can also be found in [5], [43], [59], [60], [69], and so on.

5. Sequential estimation

Let $X_1, X_2, \ldots$ be independent identically distributed p-dimensional random vectors with p.d.f. f defined on a probability space $(\Omega, \mathscr{F}, P)$. Let $\{\mathscr{F}_n, n \geq 1\}$ be an increasing sequence of sub σ-fields of $\mathscr{F}$. A random variable N taking values in $\{1, 2, \ldots, \infty\}$ is called a $(\mathscr{F}_n\text{-})$ stopping time if $\{N \leq n\} = \{\omega \in \Omega; N(\omega) \leq n\} \in \mathscr{F}_n$ for $n = 1, 2, \ldots$. By the way, there are frequent situations where the sample sizes are random. For example, when we observe customers coming to a store in time $(0, t]$ and want to estimate some numerical quantity, the number of customers may be regarded as a random variable, so that the sample size is random. In this section we shall deal with estimation of probability density functions with random sample sizes like this. In what follows, let $\mathscr{F}_n$ be the σ-field generated by $X_1, \ldots, X_n$. When N is a stopping time and $f_n(x)$ is an estimator of $f(x)$ based on a sample of fixed size n, we define the sequential estimator $f_N(x)$ by

$$f_N(x) = \begin{cases} f_n(x) & \text{if } N = n, \\ \infty & \text{if } N = \infty. \end{cases}$$

The sequential estimator $f_N(x)$ seems to have been first treated by Srivastava [78]. Davies and Wegman [13] have proposed stopping times and discussed sequential estimators and moments of the stopping times. It was on the basis of the ideas of Chow and Robbins [10], Farrell [25], and Sen and Ghosh [76], that Carroll [8] introduced two types of stopping times concerning confidence intervals of $f(x)$ of fixed width $2d$; namely, type I and type II. Wegman and Davies [91] considered another stopping time. Stute [82] considered a stopping time of type II with the RPE, and Isogai [36] discussed that of type I with the IE(a). Koronacki and Wertz [50] have introduced a stopping time such that the probability that the integrated squared error of the WWYE does not exceed some prescribed value is not less than $1 - \alpha$. In this section we shall consider the problem of sequential estimation for $f(x)$

with stopping times of type I in the one-dimensional case [36]. In what follows, let us suppose that K is nonnegative and satisfies (K2), and that $\{h_n\}$ fulfills the following conditions:

$$h_n \downarrow 0, \quad nh_n \uparrow \infty, \quad \frac{h_n}{h_{n+1}} \to 1 \ (n \to \infty), \quad \sum_{n=1}^{\infty}(n^2 h_n)^{-1} < \infty,$$

$$\lim_{n \to \infty} \frac{\log n}{(nh_n)^{1/2}} = 0, \quad \lim_{n \to \infty} nh_n^3 = 0, \quad \sum_{j=1}^{n} j^{a-1} h_j = o(n^a h_n),$$

and for given $a \in (0,1]$ there exists a constant $\beta \in (0,1)$ such that

$$\lim_{n \to \infty} n^{1-2a} h_n \sum_{j=1}^{n} j^{2(a-1)} h_j^{-1} = \beta.$$

Now let any $\alpha \in (0,1)$ and $d > 0$ be given. Here Φ denotes the standard normal distribution function, and let us set

(5.1) $$B = a^2 \beta \int K^2(y) \, dy, \quad b = B^{1/2} \Phi^{-1}\left(1 - \frac{\alpha}{2}\right).$$

Let $f_n(x)$ be the recursive estimator IE(a) ($a \in [2/3, 1]$), and let x satisfy $f(x) > 0$. In order to give a confidence interval of $f(x)$ of fixed width $2d$ with asymptotic confidence level $1 - \alpha$ for $d > 0$ sufficiently small, the stopping time $N_1(d) = N_1(d, x, \text{IE}(a))$ is defined as follows: if there exists a smallest positive integer n such that

(5.2) $$nh_n(d/b)^2 \geq f_n(x) > 0,$$

then $N_1(d) = n$, and if no such n exists, then $N_1(d) = \infty$. Let $n(d)$ be the smallest positive integer n such that $nh_n(d/b)^2 \geq f(x)$. Then, as a confidence interval of $f(x)$ of width $2d$, we consider $I_{N_1(d)}(x) = [f_{N_1(d)}(x) - d, f_{N_1(d)}(x) + d]$. First, as a result on the stopping time $N_1(d)$, we obtain the following:

THEOREM 5.1. *Let x be any point of continuity of any f. Then we have that $P\{N_1(d) < \infty\} = 1$ for all $d > 0$, and that*

(5.3) $$\lim_{d \to +0} N_1(d) = \infty \ \text{a.s.}, \quad \lim_{d \to +0} \frac{N_1(d) h_{N_1(d)} d^2}{b^2 f(x)} = 1 \ \text{a.s.},$$

$$\lim_{d \to +0} \frac{N_1(d) h_{N_1(d)}}{n(d) h_{n(d)}} = 1 \ \text{a.s.}$$

Moreover, $f_{N_1(d)}$ is strongly pointwise consistent.

Next, by using (1.1), Theorem 3.10, and a result on the limit theorem in the case where indices are random variables ([3]), the asymptotic normality of the sequential estimator $f_{N_1(d)}(x)$ is obtained as follows.

THEOREM 5.2. *Suppose that f is differentiable on R and f' is bounded. Assume further that*

(5.4) $$\frac{N_1(d)}{n(d)} \to 1 \quad (d \to +0) \ \text{in prob.}$$

Then for every x $(N_1(d) h_{N_1(d)})^{1/2} (f_{N_1(d)}(x) - f(x))$ converges in law to the normal distribution with mean 0 and variance $Bf(x)$ as $d \to +0$.

The asymptotic normality of the sequential estimator above is discussed under a more general condition than (5.4) in [46]. Using Theorem 5.2, we shall show below that the confidence interval $I_{N_1(d)}(x)$ is asymptotically consistent in the sense of Chow and Robbins [10].

COROLLARY 5.1. *Under the conditions of Theorem 5.2 we have*

(5.5) $$\lim_{d \to +0} P\{f(x) \in I_{N_1(d)}(x)\} = 1 - \alpha.$$

We consider here

(5.6) $$h_n = n^{-r}, \quad \tfrac{1}{3} < r < a.$$

As (5.3) implies (5.4), the following result can be obtained.

COROLLARY 5.2. *Suppose that f is differentiable on R and f' is bounded. Then under (5.6), for every x*

$$\frac{(1-r)bf(x)}{B^{1/2}dg(d)}(N_1(d) - g(d))$$

converges in law to the standard normal distribution as $d \to +0$, where $g(d) = (b^2 f(x)/d^2)^{1/(1-r)}$.

Let us next consider the rate of convergence of (5.5). We shall treat the WWYE as f_n, and consider $h_n = n^{-r}$, $1/5 < r < 1$. By a slight modification of the condition $f_n(x) > 0$ in (5.2), the stopping time $N_2(d) = N_2(d, x, WWYE)$ is defined as follows: if there exists a smallest positive integer n such that

$$n^{1-r}(d/b)^2 \geq f_n(x) + \frac{1}{n},$$

then $N_2(d) = n$, and if no such n exists, then $N_2(d) = \infty$. By using a result on the rates of convergence in the central limit theorem with random indices [71], we can obtain the following result [41].

THEOREM 5.3. *Suppose that f is twice differentiable on R and f'' is bounded. Then we have that $P\{N_2(d) < \infty\} = 1$ for all x and all $d > 0$, and that*

(5.7) $$P\{f(x) \in I_{N_2(d)}(x)\} = 1 - \alpha + o(d^\eta) \quad (d \to +0),$$

where

(5.8) $$\eta = \min\left\{\frac{r}{2}, \frac{2(1-r)}{5(2-r)}, \frac{5r-1}{1-r}\right\}.$$

REMARK 5.1. η of (5.8) attains the maximum $(7 - \sqrt{29})/10$ at $r = r_0 = (7 - \sqrt{29})/5$. In Remark 3.3, for $h_n = n^{-r}$ the asymptotically optimal choice of r is $r = 1/5$. Thus, the closer r is to $1/5$, the slower the speed of convergence of (5.7) becomes.

By proving the uniform integrability of $\{N_2(d)h_{N_2(d)}d^2, d > 0\}$, we obtain the following result on the moments of the stopping time $N_2(d)$ [42].

THEOREM 5.4. *Suppose that f is bounded and continuous on R. Then for every x we have*

$$N_2(d)h_{N_2(d)}/(b^2 f(x)d^{-2}) \to 1 \quad a.s. \ (d \to +0)$$

and

$$E(N_2(d)h_{N_2(d)})/(b^2 f(x)d^{-2}) \to 1 \quad (d \to +0).$$

Further, let $h_n = n^{-r}$ $(0 < r < 1)$, and let q be any positive integer. Then we have

$$(N_2(d))^q/(b^2 f(x)d^{-2})^{q/(1-r)} \to 1 \quad a.s. \ (d \to +0)$$

and

$$E\{(N_2(d))^q\}/(b^2 f(x)d^{-2})^{q/(1-r)} \to 1 \quad (d \to +0).$$

Here we shall present the stopping time proposed by Koronacki and Wertz ([**50**]). Further, suppose that K is nonincreasing on $[0, \infty)$, and $h_n = cn^{-r}$ ($2/5 < r < 1$, $c > 0$ is a constant), and that there exists the limit $0 < w < \infty$ such that

$$w = \lim_{n \to \infty} \frac{1}{n} \sum_{j=1}^{n} h_j^{-1} \int \left\{ \int K(z)K\left(z + \frac{y}{h_n}\right) dz \int K(t)K\left(t + \frac{y}{h_j}\right) dt \right\} dy.$$

Let $\{\xi_n, n \geq 1\}$ be a sequence of estimators of $\int f^2(x) dx$. Then we define the stopping time $N(\varepsilon)$ as follows: for any given $\alpha \in (0,1)$, positive integer n_0, and $\varepsilon > 0$, if there exists a smallest positive integer $n \geq n_0$ such that

$$(h_n \xi_n)^{1/2} \leq \frac{nh_n\varepsilon - B}{2(2+r)^{-1/2}w^{1/2}\Phi^{-1}(1-\alpha)},$$

then $N(\varepsilon) = n$, and if no such n exists, then $N(\varepsilon) = \infty$. Here B is as given by (5.1).

THEOREM 5.5. *Suppose that f is twice continuously differentiable and that f, f', f'' are bounded. Let f_n be the WWYE. If $\xi_n \to \int f^2(x) dx$ a.s. $(n \to \infty)$, then it follows that*

$$\lim_{\varepsilon \to +0} P\left\{ \int (f_{N(\varepsilon)}(x) - f(x))^2 dx \leq \varepsilon \right\} = 1 - \alpha.$$

Finally, we mention the definition of stopping times of type II. Let $I_n(x)$ be a confidence interval of $f(x)$ constructed by using a sample of size n, and let $|I_n(x)|$ denote the length of the interval $I_n(x)$. Then the stopping time $N(d)$ of type II is defined as follows: if there exists a smallest positive integer n such that $|I_n(x)| \leq 2d$, then $N(d) = n$, and if no such n exists, then $N(d) = \infty$.

6. Estimation of regression functions

Let $Z = (X, Y)$ and $\{Z_n = (X_n, Y_n), n \geq 1\}$ be independent identically distributed random vectors taking values in $R^p \times R$, and let the regression function $m(x)$ of Y on $X = x$ be defined by $m(x) = E(Y|X = x)$, whose functional form is assumed to be unknown. The problem of estimating nonparametric regression functions in the stochastic design model is to estimate $m(x)$ based on $Z_1, \ldots, Z_n$ (see [**9**], [**12**], [**24**],

[34], [68], etc.). Let $W_{nj}(X) = W_{nj}(X; X_1, \ldots, X_n)$ $(1 \leq j \leq n)$ be weight functions. Then as an estimator of $m(x)$, Stone [80] considered

$$m_n(x) = \sum_{j=1}^{n} W_{nj}(x) Y_j$$

and gave sufficient conditions on $W_n = \{W_{nj}\}$ for $m_n(x)$ to be consistent. We note that in the one-dimensional case $(p = 1)$, by using a kernel K and a sequence $\{h_n, n \geq 1\}$ of positive numbers with approaching zero, Nadaraya [57] and Watson [89] proposed the following nonrecursive estimator:

$$W_{nj}(x) = \frac{K((x - X_j)h_n^{-1})}{\sum_{i=1}^{n} K((x - X_i)h_n^{-1})}.$$

With regard to this estimator, the reader is referred to [22], [31], [33], [61], and [63] for example. On the other hand, when we set

$$W_{nj}(x) = \frac{h_j^{-p} K((x - X_j)h_j^{-1})}{\sum_{i=1}^{n} h_i^{-p} K((x - X_i)h_i^{-1})},$$

$m_n(x)$ becomes a recursive estimator. The statistical properties of the estimator have been discussed in [2], [20], [30], [38], [51], [52], and so on. The study on sequential estimation for the estimator has also been done ([1], [40], [44], [72], etc.). Moreover, the study for the fixed design model has also been developed. This is as follows: let $g(x)$ $(x \in R^p)$ be an unknown function. Then $g(x)$ is to be estimated on the basis of a sample $(x_1, Y_1), \ldots, (x_n, Y_n)$ which satisfies $Y_i = g(x_i) + Z_i$ $(i = 1, \ldots, n)$ $(Z_1, \ldots, Z_n$ are independent identically distributed random variables with mean 0, and $x_1, \ldots, x_n$ are given real numbers) ([28], [29], [45], [65], etc.).

7. Conclusion

So far, we have explained the recursive estimators of nonparametric probability density functions obtained by the kernel method, their statistical properties, and related problems. With regard to the RPE, a great number of interesting results have been obtained, but we hardly discussed them here. I hope that in the future studies of the estimators obtained by the kernel method will be developed in various areas. In the end, I think that higher order asymptotic theory on the basis of asymptotic expansions will also be developed on the problem of estimating nonparametric probability density functions.

References

1. M. Aerts and J. C. Geertsema, *Bound length confidence intervals in nonparametric regression*, Sequential Anal. **9** (1990), 171–192.
2. I. A. Ahmad and P. E. Lin, *Nonparametric sequential estimation of a multiple regression function*, Bull. Math. Statist. **17** (1976), 63–75.
3. F. J. Anscombe, *Large-sample theory of sequential estimation*, Proc. Cambridge Philos. Soc. **48** (1952), 600–607.
4. G. Banon, *Sur un estimateur non paramétrique de la densité de probabilité*, Rev. Statist. Appl. **24** (1976), 61–73.
5. ———, *Nonparametric identification for diffusion processes*, SIAM J. Control Optim. **16** (1978), 380–395.

6. P. Bickel and M. Rosenblatt, *On some global measures of the deviations of density function estimates*, Ann. Statist. **6** (1973), 1071–1095.
7. T. Cacoullos, *Estimation of a multivariate density*, Ann. Inst. Statist. Math. **18** (1966), 179–189.
8. R. J. Carroll, *On sequential density estimation*, Z. Wahrsch. Verw. Gebiete **36** (1976), 137–151.
9. P. E. Cheng, *Applications of kernel regression estimation: A survey*, Comm. Statist. Theory Methods A **19** (1990), 4103–4134.
10. Y. S. Chow and H. Robbins, *On the asymptotic theory of fixed-width sequential confidence intervals for the mean*, Ann. Math. Statist. **36** (1965), 457–462.
11. D. B. H. Cline, *Optimal kernel estimation of densities*, Ann. Inst. Statist. Math. **42** (1990), 287–303.
12. G. Collomb, *Nonparametric regression: An up-to-date bibliography*, Statistics **16** (1985), 309–324.
13. H. I. Davies and E. J. Wegman, *Sequential nonparametric density estimation*, IEEE Trans. Inform. Theory **IT-21** (1975), 619–628.
14. P. Deheuvels, *Conditions nécessaires et suffisantes de convergence ponctuelle presque sûre et uniforme presque sûre des estimateurs de la densité*, C. R. Acad. Sci. Paris **278** (1974), 1217–1220.
15. _____, *Estimation sequentielle de la densité*, Contrib. Probab. Est. Mat. Ens. Mat. Anal., Univ. of Granada, 1979, pp. 156–169.
16. L. Devroye, *On the pointwise and the integral convergence of recursive kernel estimates of probability densities*, Utilitas Math. **15** (1979), 113–128.
17. _____, *A course in density estimation*, Birkhäuser, Boston, Mass., 1987.
18. _____, *The kernel estimate is relatively stable*, Probab. Theory Related Fields **77** (1988), 521–536.
19. _____, *Asymptotic performance bounds for the kernel estimate*, Ann. Statist. **16** (1988), 1162–1179.
20. L. Devroye and T. J. Wagner, *On the L_1 convergence of kernel estimators of regression functions with applications in discrimination*, Z. Wahrsch. Verw. Gebiete **51** (1980), 15–25.
21. L. Devroye and L. Györfi, *Nonparametric density estimation: The L_1 view*, Wiley, New York, 1985.
22. L. Devroye and A. Krzyżak, *An equivalence theorem for L_1 convergence of the kernel regression estimate*, J. Statist. Plann. Inference **23** (1989), 71–82.
23. V. A. Epanechnikov, *Nonparametric estimation of a multivariate probability density*, Theory Probab. Appl. **14** (1969), 153–158.
24. R. L Eubank, *Spline smoothing and nonparametric regression*, Marcel Dekker, New York and Basel, 1988.
25. R. H. Farrell, *Bounded length confidence intervals for the p-point of a distribution function. III*, Ann. Math. Statist. **37** (1966), 586–592.
26. K. Fukunaga and L. D. Hostetler, *The estimation of the gradient of a density function, with applications in pattern recognition*, IEEE Trans. Inform. Theory **IT-21** (1975), 32–40.
27. M. J. Fryer, *A review of some nonparametric methods of density estimation*, J. Inst. Math. Appl. **20** (1977), 335–354.
28. A. A. Georgiev, *Consistent nonparametric multiple regression: The fixed design case*, J. Multivariate Anal. **25** (1988), 100–110.
29. A. A. Georgiev and W. Greblicki, *Nonparametric function recovering from noisy observations*, J. Statist. Plann. Inference **13** (1986), 1–14.
30. W. Greblicki and M. Pawlak, *Necessary and sufficient consistency conditions for a recursive kernel regression estimate*, J. Multivariate Anal. **23** (1987), 67–76.
31. L. Györfi, *Recent results on nonparametric regression estimate and multiple classification*, Problems Control Inform. Theory **10** (1981), 43–52.
32. _____, *Strong consistent density estimate from ergodic sample*, J. Multivariate Anal. **11** (1981), 81–84.
33. P. Hall, *On iterated logarithm laws for linear arrays and nonparametric regression estimator*, Ann. Probab. **19** (1991), 740–757.
34. W. Härdle, *Applied nonparametric regression*, Cambridge Univ. Press, 1990.
35. E. Isogai, *Strong consistency and optimality of a sequential density estimator*, Bull. Math. Statist. **19** (1980), 55–69.
36. _____, *Stopping rules for sequential density estimation*, Bull. Math. Statist. **19** (1981), 53–67.
37. _____, *Strong uniform consistency of recursive kernel density estimators*, Sci. Rep. Niigata Univ. Ser. A **18** (1982), 15–27.
38. _____, *A class of nonparametric recursive estimators of a multiple regression function*, Bull. Inform. Cybernet. **20** (1983), 33–44.
39. _____, *Joint asymptotic normality of nonparametric recursive density estimators at a finite number of distinct points*, J. Japan Statist. Soc. **14** (1984), 125–135.
40. _____, *Asymptotic consistency of fixed-width sequential confidence intervals for a multiple regression function*, Ann. Inst. Statist. Math. **38** (1986), 69–83.

41. _____, *The convergence rate of fixed-width sequential confidence intervals for a probability density function*, Sequential Anal. **6** (1987), 55–69.
42. _____, *A note on sequential density estimation*, Sequential Anal. **7** (1988), 11–21.
43. _____, *Nonparametric recursive estimation in stationary Markov processes*, Comm. Statist. Theory Methods A **18** (1989), 1309–1323.
44. _____, *Nonparametric probability density estimation using recursive kernel estimators*, Doctoral thesis, Univ. of Tsukuba, 1989.
45. _____, *Nonparametric estimation of a regression function by delta sequences*, Ann. Inst. Statist. Math. **42** (1990), 699–708.
46. _____, *A note on the asymptotic normality of sequential density estimators*, Yokohama Math. J. **39** (1992), 115–124.
47. _____, *A Berry-Esseen type bound for recursive estimators of a density and its derivatives*, J. Statist. Plann. Inference (to appear).
48. A. Kogure, *Asymptotically optimal cells for a histogram*, Ann. Statist. **15** (1987), 1023–1030.
49. _____, *Data-based cell selection rules for a histogram: A review*, Sûgaku **41** (1989), 237–245.
50. J. Koronacki and W. Wertz, *A global stopping rule for recursive density estimators*, J. Statist. Plann. Inference **20** (1988), 23–39.
51. A. Krzyżak and M. Pawlak, *University consistency results for Wolverton-Wagner regression function estimate with application in discrimination*, Problems Control Inform. Theory **12** (1983), 33–42.
52. _____, *Almost everywhere convergence of a recursive kernel regression function estimate and classification*, IEEE Trans. Inform. Theory **IT–30** (1984), 91–93.
53. M. Loève, *Probability theory. I*, 4th ed., Springer-Verlag, 1977.
54. E. Masry, *Almost sure convergence of recursive density estimators for stationary mixing processes*, Statist. Probab. Lett. **5** (1987), 249–254.
55. _____, *Recursive probability density estimation for weakly dependent stationary processes*, IEEE Trans. Inform. Theory **IT–32** (1986), 254–267.
56. E. Masry and L. Györfi, *Strong consistency and rates for recursive probability density estimators of stationary processes*, J. Multivariate Anal. **22** (1987), 79–93.
57. E. A. Nadaraya, *On estimating regression*, Theory Probab. Appl. **9** (1964), 141–142.
58. _____, *Nonparametric estimation of probability densities and regression curves*, Kluwer Academic, 1989.
59. H. T. Nguyen, *Density estimation in a continuous-time stationary Markov process*, Ann. Statist. **7** (1979), 341–348.
60. _____, *Recursive nonparametric estimation in stationary Markov processes*, Publ. Inst. Statist. Univ. Paris **29** (1984), 65–84.
61. K. Noda, *Estimation of a regression function by the Parzen kernel-type density estimators*, Ann. Inst. Statist. Math. **28** (1976), 221–234.
62. E. Parzen, *On estimation of a probability density function and mode*, Ann. Math. Statist. **33** (1962), 1065–1076.
63. M. Pawlak, *On the almost everywhere properties of the kernel regression estimate*, Ann. Inst. Statist. Math. **43** (1991), 311–326.
64. B. L. S. Prakasa Rao, *Nonparametric functional estimation*, Academic Press, 1983.
65. M. B. Priestley and M. T. Chao, *Nonparametric function fitting*, J. Roy. Statist. Soc. Ser. B **34** (1972), 385–392.
66. M. Rosenblatt, *Remarks on some nonparametric estimates of a density function*, Ann. Math. Statist. **27** (1956), 832–837.
67. _____, *Curve estimates*, Ann. Math. Statist. **42** (1971), 1815–1842.
68. G. G. Roussas, *Nonparametric functional estimation and related topics*, Kluwer Academic, 1991.
69. _____, *Recursive estimation of the transition distribution function of a Markov process: Asymptotic normality*, Statist. Probab. Lett. **11** (1991), 435–447.
70. _____, *Exact rates of almost sure convergence of a recursive kernel estimate of a probability density function: Application to regression and hazard rate estimation*, J. Nonparametric Statist. **1** (1992), 171–195.
71. Z. Rychlik, *The order of approximation in the random central limit theorem*, Lecture Notes in Math., vol. 656, Springer, 1978, pp. 225–236.
72. M. Samanta, *On sequential estimation of the regression function*, Bull. Inform. Cybernet. **21** (1984), 19–27.
73. E. F. Schuster, *Note on the uniform convergence of density estimates*, Ann. Math. Statist. **41** (1970), 1347–1348.

74. D. W. Scott, *Multivariate density estimation*, Wiley, 1992.
75. A. H. Seheult and C. P. Quesenberry, *On unbiased estimation of density functions*, Ann. Math. Statist. **42** (1971), 1434–1438.
76. P. K. Sen and M. Ghosh, *On bounded length sequential confidence intervals based on one-sample rank order statistics*, Ann. Math. Statist. **42** (1971), 189–203.
77. B. W. Silverman, *Density estimation for statistics and data analysis*, Chapman and Hall, 1986.
78. R. C. Srivastava, *Estimation of probability density function based on random number of observations with applications*, Internat. Statist. Rev. **41** (1973), 77–86.
79. E. M. Stein, *Singular integrals and differentiability properties of functions*, Princeton Univ. Press, Princeton, NJ, 1970.
80. C. J. Stone, *Consistent nonparametric regression*, Ann. Statist. **5** (1977), 595–645.
81. _____, *An asymptotically optimal window selection rule for kernel density estimates*, Ann. Statist. **12** (1984), 1285–1297.
82. W. Stute, *Sequential fixed-width confidence intervals for a nonparametric density function*, Z. Wahrsch. Verw. Gebiete **62** (1983), 113–123.
83. H. Takahata, *Almost sure convergence of density estimators for weakly dependent stationary processes*, Bull. Tokyo Gakugei Univ. Ser. IV **32** (1980), 11–32.
84. K. Tanaka, *On the pattern classification problems by learning* (I), Bull. Math. Statist. **14** (1970), 31–49.
85. R. A. Tapia and J. R. Thompson, *Nonparametric probability density estimation*, John Hopkins Univ. Press, Baltimore, MD, 1978.
86. L. T. Tran, *Recursive density estimation under dependence*, IEEE Trans. Inform. Theory **35** (1989), 1103–1108.
87. _____, *Recursive kernel density estimators under a weak dependence condition*, Ann. Inst. Statist. Math. **42** (1990), 305–329.
88. M. Watanabe, *On convergences of asymptotically optimal discriminant functions for pattern classification problems*, Bull. Math. Statist. **16** (1974), 23–34.
89. G. S. Watson, *Smooth regression analysis*, Sankhyā Ser. A **26** (1964), 359–372.
90. E. J. Wegman, *Nonparametric probability density estimation*: I. *A summary of available methods*, Technometrics **14** (1972), 533–546.
91. E. J. Wegman and H. I. Davies, *Remarks on some recursive estimators of a probability density*, Ann. Statist. **7** (1979), 316–327.
92. W. Wertz, *Sequential and recursive estimators of the probability density*, Statistics **16** (1985), 277–295.
93. W. Wertz and B. Schneider, *Statistical density estimation: A bibliography*, Internat. Statist. Rev. **47** (1979), 155–175.
94. C. T. Wolverton and T. J. Wagner, *Asymptotically optimal discriminant functions for pattern classification*, IEEE Trans. Inform. Theory **IT-15** (1969), 258–265.
95. H. Yamato, *Sequential estimation of a continuous probability density function and mode*, Bull. Math. Statist. **14** (1971), 1–12.
96. K. Yoshihara, *Weakly dependent stochastic sequences and their applications*, Vol. I, Sanseido, 1992.

Translated by EIICHI ISOGAI

Recent Titles in This Series

(Continued from the front of this publication)

121 V. D. Mazurov, Yu. I. Merzlyakov, and V. A. Churkin, Editors, The Kourovka Notebook: Unsolved Problems in Group Theory
120 M. G. Kreĭn and V. A. Jakubovič, Four Papers on Ordinary Differential Equations
119 V. A. Dem′janenko et al., Twelve Papers in Algebra
118 Ju. V. Egorov et al., Sixteen Papers on Differential Equations
117 S. V. Bočkarev et al., Eight Lectures Delivered at the International Congress of Mathematicians in Helsinki, 1978
116 A. G. Kušnirenko, A. B. Katok, and V. M. Alekseev, Three Papers on Dynamical Systems
115 I. S. Belov et al., Twelve Papers in Analysis
114 M. Š. Birman and M. Z. Solomjak, Quantitative Analysis in Sobolev Imbedding Theorems and Applications to Spectral Theory
113 A. F. Lavrik et al., Twelve Papers in Logic and Algebra
112 D. A. Gudkov and G. A. Utkin, Nine Papers on Hilbert's 16th Problem
111 V. M. Adamjan et al., Nine Papers on Analysis
110 M. S. Budjanu et al., Nine Papers on Analysis
109 D. V. Anosov et al., Twenty Lectures Delivered at the International Congress of Mathematicians in Vancouver, 1974
108 Ja. L. Geronimus and Gábor Szegő, Two Papers on Special Functions
107 A. P. Mišina and L. A. Skornjakov, Abelian Groups and Modules
106 M. Ja. Antonovskiĭ, V. G. Boltjanskiĭ, and T. A. Sarymsakov, Topological Semifields and Their Applications to General Topology
105 R. A. Aleksandrjan et al., Partial Differential Equations, Proceedings of a Symposium Dedicated to Academician S. L. Sobolev
104 L. V. Ahlfors et al., Some Problems on Mathematics and Mechanics, On the Occasion of the Seventieth Birthday of Academician M. A. Lavrent′ev
103 M. S. Brodskiĭ et al., Nine Papers in Analysis
102 M. S. Budjanu et al., Ten Papers in Analysis
101 B. M. Levitan, V. A. Marčenko, and B. L. Roždestvenskiĭ, Six Papers in Analysis
100 G. S. Ceĭtin et al., Fourteen Papers on Logic, Geometry, Topology and Algebra
99 G. S. Ceĭtin et al., Five Papers on Logic and Foundations
98 G. S. Ceĭtin et al., Five Papers on Logic and Foundations
97 B. M. Budak et al., Eleven Papers on Logic, Algebra, Analysis and Topology
96 N. D. Filippov et al., Ten Papers on Algebra and Functional Analysis
95 V. M. Adamjan et al., Eleven Papers in Analysis
94 V. A. Baranskiĭ et al., Sixteen Papers on Logic and Algebra
93 Ju. M. Berezanskiĭ et al., Nine Papers on Functional Analysis
92 A. M. Ančikov et al., Seventeen Papers on Topology and Differential Geometry
91 L. I. Barklon et al., Eighteen Papers on Analysis and Quantum Mechanics
90 Z. S. Agranovič et al., Thirteen Papers on Functional Analysis
89 V. M. Alekseev et al., Thirteen Papers on Differential Equations
88 I. I. Eremin et al., Twelve Papers on Real and Complex Function Theory
87 M. A. Aĭzerman et al., Sixteen Papers on Differential and Difference Equations, Functional Analysis, Games and Control
86 N. I. Ahiezer et al., Fifteen Papers on Real and Complex Functions, Series, Differential and Integral Equations
85 V. T. Fomenko et al., Twelve Papers on Functional Analysis and Geometry

(See the AMS catalog for earlier titles)